HOW TO SURVIVE A NUCLEAR ACCIDENT

To Jan, Lois, and Fred

HOW TO SURVIVE A NUCLEAR ACCIDENT

by Duncan Long

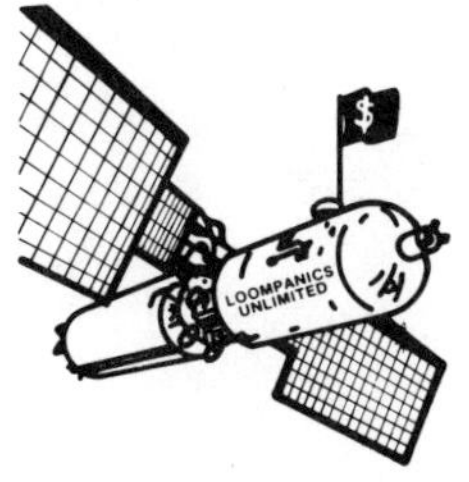

LOOMPANICS UNLIMITED

Port Townsend, Washington

HOW TO SURVIVE A NUCLEAR ACCIDENT

Printed in U.S.A.

Published by:
Loompanics Unlimited
PO Box 1197
Port Townsend, WA 98368

ISBN 0-915179-67-9
Library of Congress
Catalog Card Number 87-081612

TABLE OF CONTENTS

WHAT IS THE DANGER?

Currently fifteen percent of the world's power is being generated by nuclear energy. Because of this, most of us live, work, and/or vacation downwind from nuclear reactors most of the time. While nuclear accidents are unlikely, they do happen. And when they happen, they can be catastrophic.

In theory, nuclear materials are always handled very carefully. In reality, this isn't always the case. It is easy for workers to become casual about the moving and guarding of materials which don't always harm people instantly and which harm through rays undetectable by the senses. Thus, workers often shut down warning alarms, circumvent redundant systems, and ignore warnings. Such was the case with our modern age's latest two close squeaks with major disasters in the Three Mile Island "event" and the recent Soviet Chernobyl accident near Kiev.

While the production of new reactors has slowed following Three Mile Island and Chernobyl, new reactors are still being put "on line" in ever growing numbers. Many of these reactors may not be as safe as one might hope, especially in third world or communist countries where energy (or weapons grade materials) is often more important than individual safety and where governments are less than forthcoming about the dangers posed by their nuclear industries. Even with ultra-safe, automated, and expensive reactors, the US and Europe still find that disasters can occur which could breach the safeguards. Earthquakes (a few U.S. reactors are built on major faults and at least one reactor — according to some experts — lacks proper reinforcement), airplane crashes, shoddy workmanship, use of drugs by workers, etc., could all conceivably lead to a major nuclear reactor accident.

Also to be contended with is terrorism and war. While many experts worry about a nuclear weapon in the hands of a terrorist organization, they often fail to realize that nuclear reactors are disasters "in place" waiting for a truck bomb or sabotage to create a "dirty" accident. While modern nuclear reactors aren't capable of actually causing a nuclear explosion, as we'll see, they could release enough nuclear material to be extremely dangerous over a very wide area. Currently the nuclear industry seems to be rather lackadaisical about the risks. A Midwest firm owning a reactor was recently fined because of their insufficient security; it was possible for unauthorized "visitors" to simply walk into the reactor area!

A war could also cause reactors to release untold amounts of nuclear contaminants. In a major war, reactors would probably be among the targets that might be attacked just like conventional power plants would be. Since some nuclear plants also create materials that might be used to make nuclear bombs, such plants could also be prime military targets as well.

In a nuclear war, even if reactors aren't targeted, many experts think that the EMP (Electro-Magnetic Pulse — a lightning-like effect created by nuclear weapons) might disable the controls of a nuclear reactor so that workers would be unable to control the reactor. In such a case, a reactor might "run away" to a core melt down.

Many feel that one of the major problems with the nuclear military and industrial system is in its disposal of waste materials. While nuclear-generated electricity is "cleaner" than that of coal, the waste products produced by a reactor are much more dangerous over the long haul.

Currently, there are no widely acceptable ways of storing this waste and it is building up in many temporary storage areas. (Perhaps the worst leak, to date, from such a storage area occurred in 1971 at Monticello, MN, when 55,000 gallons of radioactive water escaped from a storage area and flowed into the Mississippi River. This material found its way into the water supply of St. Paul, where higher levels of radioactivity were recorded. While the level of radioactivity was too low to create short-term risks, undoubtedly there was a slight increase in long-term risk.)

In the US, the military has been quite lax in its disposal of waste products. In the past, the US military dumped radioactive waste at sea. Unfortunately, less than accurate records were kept of much of this dumping and some of the canisters have been found in shrimp fishing areas off California. Due to the fact that the radioactive materials involved have a much longer life than the canisters they are contained in, it would seem to be only a matter of time before such material finds its way into our food chain.

Perhaps the worst dumping at sea occurs in Britain where the nuclear industry is allowed to dump plutonium and other wastes directly into the ocean. While in theory the waste sinks in areas of the sea where it will never again be seen, in fact, much of it washes back to shore. In some areas it is hazardous to a person's long-term health to stroll along the beach; any dust stirred up by the walk can contain plutonium!

While most high-level nuclear wastes (those producing excessive amounts of radiation) are carefully stored and monitored in the US, low-level waste isn't.

Large amounts of nuclear material are used in industry, agriculture, and by medical institutions; much of this material is low-level waste and the disposal and distribution of such materials is rather casual. (For example, liquid radioactive wastes were recently transported across the US to Washington State in leaking containers. While the material was theoretically spread out enough by the long trip to pose little danger, those who might have walked through a pool of the material at a gas station or rest stop might have become contaminated enough to have health problems in later years.)

Currently this low-level waste is simply being placed in landfills without any sort of containment arrangement other than the porous earth around the material and containers.

Small amounts of nuclear materials are also disposed of by citizens in landfills and garbage dumps. This radioactive material comes from such things as glow-in-the-dark watches, smoke detectors, etc. Perhaps the worst of these materials come from "antique" watches, table ware, and other objects made before people were aware of the dangers from radioactive materials. While most of these materials pose little danger, there is always the chance that particles of such radioactive waste might get into water supplies or the air. While the risks are small percentage-wise, the person ingesting any such material might face serious health problems. Radioactive waste has become a sort of game of national Russian Roulette.

As of yet, there does not appear to be a complete solution to the processing of nuclear wastes. While a number of methods, including burial in geological formations, solidification, vitrification, and even rocketing the material into the sun have been proposed, none have proved to be both practical and stable over the thousands of years during which much of the material will be dangerous. (A Nuclear Regulatory Commission "think tank" has even proposed that a sort of nuclear priesthood be created which would guard the wastes from retrieval during the changes of civilization or any neo-dark age which might occur during the next several thousand years that the material will remain dangerous.)

Another potential problem is that currently the Nuclear Regulatory Agency allows nuclear materials to be carried on commercial airplane flights. In fact, it is illegal for commercial carriers to refuse to transport the material if they have room to carry it. That means, next time you're on a plane, you may be close to dangerous canisters of nuclear materials. Likewise, a nuclear disaster may be just a plane crash away from your home. (Added to the list are military planes with nuclear weapons. While the danger of a nuclear explosion from such a crash is small, the release of nuclear materials is possible.)

Military planes carrying nuclear weapons also pose some risk. While accidental dropping of nuclear bombs are rare, they do occasionally occur. For example, a US Air Force bomber accidentally released a 42,000-pound, 10-megaton bomb in 1957 when flying over New Mexico. While the Mark 17 bomb didn't create a nuclear explosion, some of the conventional explosives in its trigger did go off, releasing some nuclear materials from the bomb's nuclear core. Because there was no actual nuclear detonation, the US Defense Department recently said that this incident "confirms the efficacy of the safety devices." While the chances of such an accident are small, disaster could fall out of the air at any moment.

Or out of space.

A number of Soviet satellites have been placed in orbit which use nuclear reactors for power. As the Canadians have discovered, such satellites can fall back to Earth and leave miles of contaminated fragments over the water and soil.

Finally a lot of nuclear material seems to get "lost." Whether this is through bookkeeping errors, theft, or illegal dumping is unknown in the industry (and is something rarely mentioned to the public). But even if the radioactive material is unaccounted for, the material is still dangerous. It is probable that land fills containing much of this material are sitting somewhere, like hidden time bombs, waiting to go off. Worse yet, some of this material may be finding its way into the hands of terrorists.

The good news is that the nuclear industry has a fairly good track record for safety thus far. The bad news is that just one mistake could spell the deaths of hundreds or thousands of unsuspecting people and ruin the health of untold numbers of others.

What about government safeguards?

In capitalist countries, there is a tendency by private industry to minimize safeguards wherever possible to maximize profits. Money is what makes things go in a capitalist society and — as things stand now — little monetary reward goes to companies that take extra steps to protect those around their plants.

Likewise, in socialist or communist countries, the well-being of the masses is often more important than that of individuals. Lack of power can mean slow economic growth and lower the standard of living. Cheap power doesn't remain cheap if you pile nuclear reactors with redundant safeguards. Often the choice is made to take a few risks to help meet long-range goals of growth. (Following the 1986 Soviet report to the International Atomic Energy

Agency at Vienna, Austria, about the Chernobyl disaster, the Russians stated that they would continue to aggressively pursue the development of nuclear energy because combustion of fossil fuels posed such a great danger to the environment!)

Face saving comes into play, too, in communist countries. Nuclear reactors are status symbols of progress as well as the ability to produce nuclear weapons. And accidents take away from the status; the USSR would never have bothered to mention Chernobyl if Sweden, Denmark, and Finland hadn't noted the high levels of radiation — up to 100 times the normal level — generated by the accident. Only after the West had pieced together the location of the disaster by studying wind charts did the USSR finally admit to a problem — nearly a week after the event.

A lot of power is lost in "piping" electrical energy over wires. Therefore, in both the East and West, power plants are often located near the population centers which will be demanding large amounts of power. While such a site is wise from a power-saving standpoint, it makes little sense from a safety standpoint. Many reactors couldn't be more poorly placed if the designers had been trying to maximize the damage that might result from an accident.

In both the East and West, governments also have a vested interest in keeping the nuclear power plants operating. Lack of energy can mean a change in government; in an age of OPEC and dwindling supplies, this is no small consideration. Solar energy, wind energy, tidal energy, and geo-thermal energy have all proven to be too expensive or unreliable to be workable in large-scale systems which can be controlled by a power company. While government inspectors are supposed to safeguard citizens, they are also under pressure to keep plants open so that power can be generated. Often, carelessness and lack of safeguards which would normally close a plant may be overlooked in order to avoid blackouts and social unrest.

On the local level, things aren't a lot better. Preparing for a mass evacuation in the event of an accident takes time and costs money. And ordering such an evacuation means loss of time and money as workers leave their jobs and homes are left untended. Ordering an evacuation when it isn't needed is also a major political embarrassment. So local officials are under a lot of pressure to wait until the last minute to do anything. That could be too late. (In both Three Mile Island and Chernobyl, public officials waited too long before ordering any evacuation of the population. Chernobyl wasn't evacuated until 36 hours after the accident; had it been worse, it is probable that thousands more would have died while those in charge worried about looking bad.)

Likewise, to save money, police units and civil defense workers often don't have the radiation detection equipment needed to deal with small or large accidents. Since radiation is undetectable to the senses, in an emergency it is often like the blind leading the blind when officials try to deal with radiation and make decisions as to what should be done.

If government regulators and the nuclear industry itself all have vested interests that make your safety less than ideal, what can you do to protect yourself?

You could just act like there is nothing to worry about. That's what many did at Three Mile Island right up to the day that the city next to it was ordered to be evacuated. The accident might just as easily have gone the other way and killed many of those who "knew" they were safe.

A better strategy is to take a few simple precautions and buy some relatively inexpensive equipment. You *can* improve *your* chances of surviving a nuclear accident — whether it's a minor or major accident. As we'll see, there are a number of simple steps you can take to greatly improve your survival chances. A little time and thought will enable you to cope with any dangers which may occur in your area. Too, knowing the dangers will take a lot of the fear and mystery away from them and help you to cope more realistically with them.

The bottom line is this: preparing to avoid receiving dangerous exposures to radiation or even being killed is, for the most part, up to you. You must prepare to protect yourself. Depending on industry or politicians may cost you your health or life.

Let's look at what you might be facing in the way of nuclear accidents.

THE CHINA SYNDROME AND OTHER GOODIES

Nuclear accidents become dangerous to the general public when radioactive materials somehow escape to remain in the environment. Here they are hard to detect and contamination of water, food, etc., goes undetected. When wastes are spilled from containers or radioactive gas or dust is released from industrial accidents, the spread of dangerous materials isn't hard to imagine. Nuclear contaminants can be tracked on the bottom of shoes, on automobile tires, or even spread to other continents by migrating birds and insects.

(This includes wastes which can be extremely dangerous. In 1957, near Kyshtyn, USSR, a nuclear explosion apparently occurred when nuclear wastes which were being stored underground reached a critical mass and started a run-away chain reaction. Although it is known that a major evacuation of the area took place and that several small villages "vanished" from Soviet maps, the USSR has never admitted to the accident. The numbers killed and injured are unknown in the West.)

Radioactive wastes don't get into the environment only by accident.

Recently large amounts of radioactive steel were found in buildings where the bars were being used for structural reinforcement. The steel had come from old nuclear containment buildings and had been sold as scrap by unscrupulous businessmen in the US. Not only did the steel end up in a lot of concrete buildings in Mexico, some was sold back to the US. The problem was only discovered when a truck carrying the bars entered a nuclear industrial area and the alarms went off — this time when something entered rather than left!

Likewise, the "tailings" (waste material left over when uranium is refined) have often been used to make concrete for buildings. Many homes and buildings in Colorado are mildly radioactive.

While the short-term health hazards from living in a home made with tailings or working in a building made of radioactive steel are small, the statistics aren't in yet on long-term health risks. In effect, those living and working in such structures are guinea pigs in studies of long-term effects of low-level exposures. Terrorists don't need to make a nuclear bomb or blow up a reactor to contaminate an area with nuclear material. All that needs to be done is to place stolen materials around a conventional explosive. Placed on a high building or dropped from a plane, the explosion would powder the material and the winds would carry it over a large area. While the contamination would be low-level for the most part, it would be great enough to create severe problems and force an evacuation of the area.

However, all these risks pale in comparison to the worst events which can occur when a nuclear reactor gets out of control.

Such events are *not* the same as the nuclear reaction which occurs when an atomic weapon explodes. A reactor uses fission products (many of which could be used to make a bomb); but in a reactor the material is not placed in dense enough groupings to create a run-away chain reaction.

So there would be no nuclear explosion from a run-away reactor. Large amounts of radioactive material can be released in several other ways, however.

One way is through a fire like that of the Chernobyl reactor. While most reactors don't use graphite blocks to moderate chain reactions, there are a few that do even in the US (these reactors are ideally suited to producing nuclear weapons grade material — a fact ignored by much of the press in the Chernobyl accident).

The decay of uranium in a nuclear reaction creates high-velocity neutrons which must be slowed down, or moderated, in order to enhance the chain reaction creating energy in a reactor. One way of doing this is to use graphite blocks which moderate the neutrons. The problem with graphite is that the carbon atoms in it do not always return to their original positions in the graphite crystal. This "misplacement" creates a store of energy called "Wigner energy." This energy creates heat which gradually builds up if the reactor isn't shut down periodically to "purge" the reactor of this energy. The heat build-up can result in a high temperature which can finally lead to a fire in the graphite moderators.

Because of this or other accidents, the graphite can catch fire. As the high temperatures created by the fire continue, up to 25 to 70 percent of the radioactive material in the reactor can be vaporized and wafted up into the air to be carried over long distances by the wind. A second problem can also occur when the high heat is created by the burning graphite. The temperature can cause water — often used to cool the core — to break down into an explosive mixture of hydrogen and oxygen.

Apparently graphite reactors have a tendency to do such things; Chernobyl was not the first graphite reactor to burn. In October, 1956, the Windscale reactor near Liverpool, England, had its graphite cubes ignite. The resulting fire and hydrogen/oxygen explosion contaminated a large area of farming land and officials banned the sale of milk produced by cows in the area. British officials believe that at least 39 people have already died of cancer which can be linked to the accident.

Many in the US think that the Chernobyl reactors had no containment shells built around them. However, researchers for the US Nuclear Regulatory Commission claim that Chernobyl actually had two types of steel and concrete barriers built around their reactors in the form of 1-foot-thick steel walls backed by 6 to 8 feet of reinforced concrete. These and other containment structures — according to the NRC — were designed to contain pressures up to 57 psi (pounds per square inch) for the steel shell and 27 psi for the concrete shell. Since most US and European containment shells are designed to survive levels up to 55 to 65 psi, there has been some concern in the nuclear industry that containment shells might not actually contain explosions which could occur in a nuclear accident.

While non-graphite reactors are less apt to have explosive gases form in them, it is still possible for a core to create explosive gases. Hydrogen, methane, and/or carbon monoxide gas may be created during an accident. These gases are highly flammable, or even explosive, depending on their concentration. While the dangerous mix of gases can be vented from the shell without problem, if radioactive gas or dust has breached the reactor and is being held in the containment building surrounding the reactor it creates a dilemma.

While an explosion of the gases would *probably* not be powerful enough to damage the containment shell, it might create damage to many other systems within the nuclear reactor building. Those running the shelter have to either release the gas — along with the nuclear contaminants — or contain them and risk a possibly disabling explosion.

Another possibility for disaster in any reactor can occur if the cooling system and any back up cooling mechanisms in the reactor fail completely (or are shut off for some reason by workers). When the temperature goes up, the radioactive core can quickly melt its control rods so that they would become inoperable. In such a case, the reactor continues generating heat unchecked in a run-away chain reaction.

The excessive heat could create a melt down during which the radioactive core fuses into a molten mass. Some authorities feel that this heat could be great enough to actually melt the floor of the concrete containment shell and the earth below it.

The worst-case scenario of such a melt down is the *China Syndrome.* According to this as yet unproven theory, the molten core would melt its way downward deep into the earth ("clear to China," as the saying goes). The molten mass would drop into the earth until it finally quit its reaction or ran into ground water.

While the public has generally been led to believe that such events would be one-in-a-billion, the record shows that the theory has nearly been tested a number of times in the short period during which nuclear reactors have been in operation.

In addition to Three Mile Island and Chernobyl, there have been a number of other *known* melt downs (and the possibility of others in the USSR, China, etc., which have not been made public). In 1952, the Chalk River reactor in Ottawa, Canada created the first known partial melt down; in 1959, a partial melt down occurred in the Santa Susana, CA, reactor; in

1966, the Enrico Fermi Breeder Reactor, Detroit, MI, had a partial melt down. Additionally, there have been a number of other accidents where coolant systems failed momentarily allowing a heat build-up which came just short of a partial melt down. While proponents of nuclear energy can argue that reactors are very safe and have many redundant systems, opponents might reply that it appears to be only a matter of time before we find out just how bad the China Syndrome can be.

Reactors in the US aren't as dangerous as they were prior to the Three Mile Island accident. Following the crisis at Three Mile Island in 1979, the Nuclear Regulatory Commission ordered a number of changes both in procedure and in the use of instrumentation and protective equipment to prevent such an accident from happening again. (Just how effective these retrofits are remains to be seen; in the meantime, they have cost an average of $65 million dollars for each of the plants involved.)

Some plants were exempted from these changes, however. Nine plants related to the US Military or US Government research projects are managed by the Department of Energy rather than the Nuclear Regulatory Commission. Therefore, these plants don't have to meet commercial safety standards and some even lack containment shells. Many feel these plants are potential "Chernobyls." There are currently nine of these nuclear reactors in operation, including one at Hanford, WA; one at Idaho Falls, ID; four on the Savannah River, SC; and two near Oak Ridge, TN. Unfortunately the Department of Energy doesn't require containment buildings for its reactors; many feel these reactors expose the public to needless high risks if a major accident should ever occur at any of these nine reactors.

In August, 1986, in an assessment of the Hanford site by the US GAO (General Accounting Office), it was found that one of the reactors was deteriorating to the point where it could be dangerous to public safety. Cost of bringing this one reactor up to safe standards carried a $1.2 billion price tag. (In a strange move, Sen. Mark Hatfield of Oregon recently pushed legislation through his subcommittee on energy and water that *removed* $21 million dollars earmarked for renovation of the Hanford reactor; the reason: it *might* force the DOE to close the plant.)

Other factors can also have detrimental effects on the safety of a reactor.

For one thing, there is the "human factor." Not all the blame here should go on the workers (though this is where the nuclear industry and governments often like to place it). Often the over-all design is too complex, and the instruments often exhibit poor human engineering, which makes it hard to control the reactor or determine exactly what is happening inside it. While errors are made because the controls are poorly designed, technically the errors are "operator mistakes" even though the blame doesn't necessarily rest on those working in the reactor.

(A good example of poor design and bureaucratic blame took place with the Chernobyl accident of 1986. When the Soviets finally made their official report to the International Atomic Energy Agency, the blame went to human error. Valeri Legasov, the first deputy director of the Kurchatov Atomic Energy Institute described what had happened then added, "The sequence of human actions was so unlikely that the engineer (who designed the system) did not include such a scenario in his project." Days after making this statement, Russian officials admitted that at least half of the reactors which were similar to the

Chernobyl reactors had been shut down for undisclosed "repairs."

Rudolf Schulten, a West German nuclear physicist said of the Chernobyl incident, "The accident was mainly due to human error, but the reactor itself is a very old-fashioned type. The safety philosophy of this reactor would never be accepted today by any country in the Western world." (Adding insult to injury is the impression left by the Soviets that the workers somehow came up with the experiment that led to the accident on their own and acted without orders from others.)

Another problem is that workers aren't a part of the machinery running the reactor, even though designers may look at them that way. For example, most industrial accidents happen on the weekend. The reason: managers and key personnel take off the weekend while less skilled workers are stuck with the weekend shift. Problems that could be easily and quickly corrected when skilled personnel are at work may lead to a disaster.

Kennedy Maize, a senior analyst at the Union of Concerned Scientists made remarks about the Chernobyl accident that are far from comforting. He recently said that although the Soviet reactor appeared to be run "by the Marx brothers...it struck me as terrifying that this whole comedy of errors could actually have taken place." He went on to say that the pattern was "...not inconsistent with what we have seen at US plants."

Greed, graft, carelessness, and indifference can also create plants that are dangerous. If a high-tech plant is to work safely, it requires that design specifications be carefully followed during its construction. Often construction companies are either unable to do such quality work or chose to cut corners in an effort to save money. A few plants have been so poorly built in the US that they have never opened while many others have gone on line amid charges by (often fired) inspectors that the plants aren't safe.

Finally, protection and security cost money. Often precautions against vandalism or sabotage are minimal in an effort to keep costs down.

What happens when one of these problems causes a "fail safe" system to fail? What makes a run-away reactor so dangerous?

If a total melt down ("China Syndrome") ever takes place, having the molten core hit ground water would be the most likely — and most disastrous — happening. Because of the intense heat of the melt down, any ground water encountered would be vaporized on contact. The steam created could make an extremely high jet of energy which might propel much of the molten core upward with explosive force. Many experts feel this explosive action would shatter the core and force the material skyward to rain back down on the earth.

The extent of the contamination from such an event would vary according to the wind speed, size of the core, size of the particles produced by the accident, the amount of heat and molten material involved, and the amount of radioactive materials freed by the accident (undoubtedly along with a lot of unforeseen factors as well).

The pattern of contamination would be similar to that of a nuclear weapon exploded on the surface of the earth *except* that the contaminated particles would not extend nearly so high into the atmosphere and would therefore cover a much smaller area. Rather than the fallout being carried by high-altitude winds, the pattern of fallout created by a major melt down would be determined by the wind at ground level. Only very small dust particles and gas would ever reach the altitude to be spread by high-altitude winds.

Unfortunately, unlike a nuclear explosion, a melt down and steam explosion would create fallout which would remain dangerous over thousands of years rather than a few weeks.

In such an event, there would be two different areas and levels of radioactive contamination. The most dangerous contamination would be created by the large radioactive core fragments, parts of the structure of the reactor (which may have residual low-level radioactivity), and other materials which may have picked up small amounts of contamination when exposed to the molten core.

These chunks of material would vary in size from very large blocks of the structure to fine sand or dust proportions. The larger the material, the higher its wind resistance and initial steam-propelled velocity; large pieces of material would fall closer to the reactor. Small debris would have a higher initial velocity and would be displaced more by the wind; the smaller the material, the farther from the reactor it will finally fall to earth.

Thus, the most dangerous areas following a reactor disaster will be near the reactor with the danger level gradually becoming less with distance from the reactor.

This dangerous area would probably not have a radius of more than a few miles (though this is hard to predict because of the many unknowns involved in what might happen) and might even be as small as only several hundred yards. Levels of radiation might be as high as 200 to 3,000 rems per hour within this area; anyone in such an area would be facing the possibility of extreme health effects or even death in a short period of time. Rescue operations would be extremely hard to carry out in such an area.

(The explosion and fire of the Chernobyl reactor created a high-level area which, seven days after the accident, had *diminished* to 200 rems per hour according to Moscow Communist Party leader, Boris N. Yeltsin.)

As distances from the accident site increased, the amount of contamination would decrease. Within areas 2 to 10 miles from the disaster, radiation levels would be much lower than those near the reactor. Such levels would still be capable of creating serious health problems, however, and the area would have to be avoided or properly decontaminated before it could be entered for any period of time.

Within this area of high contamination, death could result in a matter of hours or months from exposure to the radiation.

(It is interesting to note that early reports made by those who had access to pictures taken by US spy satellites were talking of thousands of casualties as a result of the Chernobyl accident. This certainly might have been the case. Whether the actual "official" Soviet reports of thirty-some casualties with 40 or more on critical lists is correct or not is open to conjecture. At any rate, casualties in this area can range all over the board depending on the speed of the evacuation and the amount of nuclear material involved.)

A second level of contamination will be created by the radioactive gases and fine dust produced by the nuclear melt down. Most of this material will be airborne for some time with the finest particles never reaching the earth or only being brought down by rain or ice particles. Thus, the contamination could reach the ground very gradually or suddenly if trapped in rain or snow particles.

(One way of avoiding trouble is to stay out of rain and snow as well as puddles of rain water following a nuclear disaster.)

Depending on the many variables, levels of radiation from this material would probably be small beyond 20 to 50 miles from the center of the disaster. Levels of radiation would probably gradually decrease out to 100 to 300 miles where the only risks will be long-term health risks for those exposed to the material.

Exposure to radiation beyond several hundred miles out to several thousand miles would probably be small and would create risks only of increased likelihood of cancer, leukemia, etc. And at this distance, the time lag between exposure and symptoms would be in terms of years.

These long-term risks are small but nonetheless noticeable when large populations are involved. Months after the Chernobyl disaster, scientists in the US believe that the total number of Soviet people who will die prematurely from radiation-induced diseases (including thyroid cancer) will be somewhere between 5,000 to 24,000 with a few thinking it might even reach 70,000 premature deaths.

THE DANGERS

While most of us have grown up with a healthy (or even unhealthy) fear of radiation, it is a totally alien thing to all of us even though it is actually a part of the everyday environment. In fact, cosmic rays come from outer space and penetrate — and damage — our bodies day after day. Many areas also have a low level of background radiation from various rocks or radon gas that occurs naturally in the area.

Over a year's time, most of us receive enough gamma ray damage to equal a number of high-powered x-ray examinations. While the human body is able to repair the damage faster than it is created, it is probable that, over time, some of this radiation will cause various forms of cancer or at least increase the likelihood of its formation.

The dangerous type of radiation found in the nuclear industry is known as "ionizing radiation." Unlike many other types of radiation like radio waves, visible light, microwaves, ultraviolet light, or infrared radiation, ionizing radiation strips parts of atoms apart — creating unstable charged bits of matter called "ions." Any material which gives off ionizing radiation is called "radioactive."

In living things, these ions translate into damage to living cells. The greater the amount of ionizing radiation which you are exposed to, the greater the damage done to your body. While government agencies worldwide enjoy setting limits and thresholds for exposure to radiation to workers or the general public, recent studies suggest that there is no actual cut off point below which exposures are actually safe to receive.

In fact, the more exposure, the more damage to your body. There is no such thing as a "safe" limited exposure; there are only "less damaging" exposures. The less exposure you receive, the better for your long-term health it will probably be. Whether you're in a nuclear accident or getting X-rays at the dentist's office, minimizing any exposure you have to take will give your body a better chance to repair the damage and will reduce long-term risks of cancer and other radiation-related diseases.

Four types of ionizing radiation are created by nuclear materials. These include alpha particles, beta particles, gamma rays, and neutrons.

Alpha and beta radiation have very low power levels so that they can be stopped by thin layers of material; your skin will stop alpha ray penetration while a layer of heavy clothing, aluminum foil, or even a piece of paper will stop both beta and alpha particles. Because of the low levels of energy and penetration these two particles exhibit, alpha and beta radiation are considerably less dangerous than neutrons or gamma rays.

Alpha particles are composed of two protons and two neutrons and have a positive charge. Because of their relatively large mass and positive charge, they don't penetrate through much material before they are stopped. Beta particles appear to be identical to electrons and have a negative charge. Because of their low mass, they have more penetrating power than alpha rays.

Neutrons and gamma rays penetrate light-weight materials very easily, so only thick layers of dense material will protect a person from gamma or neutron radiation.

(X-rays are similar to gamma rays but generally are created in a beam and usually are more fully "controlled" than are gamma rays which tend to travel in random directions depending on how the isotope creating them is decaying.)

Gamma rays are generally considered to be an electromagnetic radiation because the particle creating them (if any, according to the physics theory you embrace) is so small. They are highly penetrative,so it requires a lot of dense material to shield an area from them.

Neutrons are to be found in the nucleus of every atom. Therefore, neutrons are normally safe. While they have a relatively large amount of mass, neutrons have no charge; therefore, they are able to penetrate a lot of material before being stopped by a collision with the nucleus of an atom. When they are freed by the decay of a radioactive substance, neutrons have a lot of power and can penetrate a lot of material. Neutrons are most often encountered in the nuclear industry during the fission (or splitting) of uranium isotopes which are used to fuel nuclear reactors.

Many of the normally stable elements which occur in nature are to be found in an isotope form and would also be present following a nuclear accident. These are chemically similar to their non-radioactive counterparts. Those which are normally used by the body for nutrients or which mimic other nutrients, are very dangerous since they can be incorporated into the body where they continue to give off radiation for a long period of time. Over time, even low levels of radiation would be capable of creating grave damage to the human body.

The amount of energy which is given off by a radioactive substance can be measured a number of ways. When dealing with human beings in a civil defense/nuclear disaster standpoint, the most common measurements in the US are "roentgens" and the "rem" (roentgen equivalent, man). Since radioactive accidents could involve minute amounts of radioactive materials and since the nuclear industry generally monitors units of radiation smaller than the rem, the "millirem" is often used. This is simply a thousandth of a rem. This is abbreviated as "mr."

Another unit of measurement that is becoming popular in the medical and nuclear industry is the "sievert" which is abbreviated as "sv." The sievert is a large unit so it is generally given as milli-sieverts (or thousandths of sieverts). One millirem is equal to 100 milli-sieverts.

In Europe, the measurement gaining popularity is the "Gray;" normally "CentiGrays" are used in this measurement as they are a handier size. A CentiGray is equal to one roentgen.

For practical purposes, when dealing with people and the effects of radiation on them, the CentiGray, roentgen, and rem are all basically interchangeable with the 0.01 sievert equal to any of them.

(There are two ways of recording radiation. One is to show what the rate is per hour if the amount being monitored didn't change during an hour's time. The other is the total amount of radiation. Care should be taken not to confuse the two readings as the total exposure to radiation can be quite different from one to the other.)

How much radiation exposure is safe as far as short-term risks are concerned? There is some disagreement in the field at this time, but the following chart should give you some idea of what you can stand in the way of radiation during a short period of time. Doses below 100 rems will not produce any noticeable symptoms so the chart starts at this 100 rem threshold. Doses above 1,000 rems in a short period of time will be lethal within 1 hour to 14 days even with medical help. Onset of symptoms with such a massive dosage will occur within 30 minutes and will probably include some or all of the following: diarrhea, fever, convulsion, tremor, ataxia, and lethargy. Death will occur from a circulatory collapse around the 1,000 rem threshold or from respiratory failure and brain edema as higher levels of exposure are encountered.

Whether a person survives after receiving doses in excess of 200 rems would depend on their state of health, body size, and whether or not medical help was available. Obviously procedures like blood transfusion or bone marrow transplants are going to require skilled medical help and facilities which may not be available if a large number of casualties have resulted or the accident occurs in an area which is hard to reach.

You should remember that this chart applies only to healthy adults who receive radiation in a short period of time. Over a longer period of time — months or years — your body has more time to repair itself and you can take a much larger total dosage without the ill effects shown on the chart.

A few people have a higher tolerance to radiation than is shown on the chart. Some also seem to have lower tolerances. A general rule of thumb for this is that youngsters, elderly people, or those with poor health will have lower tolerances. Prenatal exposure is especially dangerous for the unborn; exposures as low as 1 to 5 rems create a much higher incidence of cancer in later years.

Also, the information on the chart is geared to the effects of total body exposures. If exposure occurs only to some parts of the body, a person's chances of short-term symptoms might be much less. For example, if you unknowingly handled material which gave a large dose of radiation to the skin on your hands but which only emitted Beta rays which were stopped by the rest of your clothing, you would suffer intense burns on your hands but would be able to survive without other ill effects because the rest of your body isn't damaged. While this type of exposure generally doesn't happen, it is something to keep in mind.

RADIATION DOSES AND EFFECTS

DOSE:	100-200 rems	200-600 rems	600-1000 rems
VOMITING:	5-50%	50-100%	100%
DELAY TIME:	3 Hours	2 Hours	1 hour
1st SIGN:		Hematopoietic	Tissue
SYMPTOMS:	Moderate	Severe leukopenia; leukopenia purpura; hemorrhage; infection. Epilation above 300 rems	
CRITICAL PERIOD:	---------	4-5 Weeks	6 Weeks
THERAPY:	Reassurance; hematologic surveillance	Blood Transfusion; antibiotics	Bone marrow transplant
PROGNOSIS:	Excellent	Good	Guarded
CONVAL-ESCENCE:	Several weeks	1-12 Months	Very long
DEATH:	None	0-80%	80-100%
CAUSE OF DEATH	--------	Hemorrhage and/or infection	

Many of the symptoms created by "radiation sickness" are only temporary. For example, loss of hair and sterility induced by high exposure to radiation is normally only temporary. (Also, there are no known genetic changes created by radiation which are transmitted to offspring in human beings. Apparently the temporary sterility is a mechanism which prevents this.)

Radiation "sickness" is not contagious. Those who are sick should be checked and cleaned so that they don't spread contamination which might be on them but otherwise they can't cause others to become sick. However, because the body's resistance to disease can be lowered by massive exposure to radiation, victims can easily catch diseases which are contagious. Therefore, some precautions are in order when dealing with people who are ill from excessive exposure to radiation.

There are several ways to protect yourself from radioactive materials. The principles involved are quite simple but very effective.

Probably the easiest and most effective way of avoiding the dangers of radiation is to be far away from it. This common-sense approach is the simplest and often the most effective. It actually applies at any distance from miles to inches and is one of the basic "laws" of physics. Here's how the law goes: *when the distance from a source of radiation is doubled, the amount of radiation is quartered.* This is often referred to as "geometric shielding," even though it has to do with using distance to reduce the amount of radiation rather than any type of real shielding.

How does this work out?

Suppose you're in a heavily contaminated area near a major meltdown. You're receiving 1 rem per hour at one foot from a chunk of radioactive debris leaning against the outside of your house. You've managed to keep the fallout dust out of your house and — for the sake of simplicity in our example — there are no other major sources of radiation nearby. By moving 2 feet from the debris, you would reduce the amount of radiation you receive to 0.25 rem per hour. Moving to 4 feet would drop the dosage to 0.063 rem per hour.

While this method of reducing exposure may not help much in the "real world" where there might be a number of sources of radiation, it can be of use if you are forced to stay in an urban environment where there are large buildings in a heavily contaminated area. In such a case, being in a large building could be worked to your advantage by staying in the center halfway between the top floor and the ground floor. Doing so would give you geometric shielding from the radioactive debris on the ground and that on the roof of the building. Moving to the center of the building or into the hallway of a hotel would then give you geometric shielding from fallout on window ledges or coming from other buildings. A large building could afford you a lot of safety in such a scenario.

Another way of reducing the amount of radiation you receive is by "time shielding." The idea here is that, with time, radioactive materials decay and become less dangerous. This is important with isotopes with short half lives; but with many of the products which might be encountered from various types of nuclear accidents, the half lives are in hundreds or even thousands of years. Therefore, unless you're Methuselah, this isn't much of a consideration. (It is a consideration with nuclear weapons which produce fallout having a very short half-life. For a detailed look at how to survive a nuclear war —

and it is possible — see my book *Nuclear War Survival*, available from Guillory and Associates for $7.95.)

The third method of reducing exposure to radiation is truly "shielding." It's called "density shielding." This varies with the type of radiation you're dealing with. In the case of alpha and beta particles, heavy layers of clothing or other thin material is all you need in the way of protection. With the highly penetrating gamma rays and neutrons, however, you need much more material for protection.

The reduction of radiation by a heavy material varies according to the density of the substance involved and the speed, size, and charge of the radioactive particles. But it is possible to create a chart which "adjusts" the thicknesses of different materials to give an idea of how much material is needed to stop a given amount of gamma and neutron radiation.

Charts and tables are generally made to show how much material is needed to reduce the radiation to l/l0th or 1/2 of its original amount. This amount of material needed to reduce an amount of radiation by one half is usually known as its HVT (or Half-Value Thickness). These figures are especially useful for figuring safety factors for areas offering shelter from radioactive fallout created by nuclear weapons; but they can also show how much protection you might be able to find in various types of building materials that are in your environment following a nuclear disaster.

The following chart gives the approximate HVT's (they're approximate because the energy levels of neutrons can vary somewhat depending on their source).

HVT (HALF-VALUE THICKNESSES)

Material	HVT
Lead	0.3 inches
Steel (or iron)	0.7 inches
Aluminum	1.9 inches
Glass	2.1 inches
Standard concrete	2.2 inches
Firebrick (fireplace)	2.6 inches
Packed earth or bricks	3.3 inches
Sand or gravel	3.3 inches
Hollow cement ("cinder") blocks	4.4 inches
Water	4.8 inches
Ice	5.3 inches
Human body (average)	5.8 inches
Magazines (slick pages, stacked)	5.8 inches
Sheet rock (gypsum)	6.3 inches
Books or pulp magazines	7.0 inches
Hardwood (maple, oak, etc.)	7.7 inches
Pine wood	8.8 inches
Plywood (dry)	11.7 inches

Again imagine that you're in your house with a piece of radioactive debris leaning against it. This time you're getting a reading of 4 rems per hour from the source of radiation. If you'd place the thickness listed above of any of the materials between you and the source of radiation, you could drop the level to half its value or 2 rems per hour. Using another thickness of any of the materials listed above would drop it by half again to 1 rem per hour and so on.

Suppose you have some bricks sitting around the house. Piling up a one-brick-thick wall (each brick just *happens* to be 3.5 inches wide for our example) will reduce the exposure from 4 rems per hour to 2 rems. Another layer would reduce that amount by half or to 1 rem per hour. Another layer, 0.5 rem per hour and so on.

To find your total exposure (to check possible health effects from the medical chart above), you'd multiply the length of exposure by the hours you are forced to stay before being evacuated from the area. Suppose it were 6 hours before the National Guard comes in to rescue you and you've lowered the exposure to 0.5 rem per hour with three layers of bricks stacked between you and the wall. Six hours times 0.5 rem hours would give you a total exposure of 3 rems; you'd not notice the effects if you were a healthy adult. And remember that this would only be *if* you were next to the wall. You could use geometric shielding (moving farther from the source of radiation) to reduce the dose you receive even further.

Of course, in the real world, there probably would be more than one source of radiation to complicate things. But, as we'll see later, this method of reducing radiation can be used both for initial protection before or during an evacuation of an area or for cleaning up and decontamination following an accident by burying radioactive material in soil (i.e., placing earth density shielding between the material and those on the surface of the ground).

Finally, how long can you remain in a contaminated environment before you start to suffer ill effects? This is important to know if you have to engage in some of the decontamination procedures outlined later in this book or are forced to remain in a contaminated area for an extended period.

Again, there are no hard and fast figures because people vary in their tolerance to radiation and it appears that any radiation is bad in that it increases your chances for long-term health problems. The US Government has established maximum yearly rates of exposure for US citizens; these are 170 millirems for the general public with 5 rems for nuclear industry workers over 18. Additionally, the government allows a once-in-a-lifetime exposure of 25 rems during times of emergency. If, however, you are willing to risk long-term health problems later on, the following chart may be used in an emergency and would give maximum doses which wouldn't affect a healthy adult's short-term health.

It is important that you keep careful track of total radiation exposure rather than "guesstimating," in order to avoid radiation sickness. While you could probably take much more radiation without ill effects, limiting your exposure to a total of 15 rems per week would be wise. Again, the less radiation you receive, the less chance you have of suffering ill effects from it.

Time must also be allowed for decontamination (as outlined later) following your time in a contaminated area.

What about countering the effects of any low-level radioactive gases or dust that you may ingest or be exposed to? While it is possible to evacuate the most dangerous areas directly surrounding the site of a nuclear accident, it would be all but impossible to evacuate the areas which receive low levels of radiation. And while these low levels don't pose immediate dangers, they do pose long-term risks.

We'll look at removing the contamination from your skin and clothing as well as how to process food and purify water (even in a heavily contaminated area) in a later chapter. These methods will minimize the chances of spreading much of the contamination to your food supply or breathing radioactive materials. Also covered later will be methods of using masks to reduce the radioactive material you breathe.

SAFE TIME IN RADIOACTIVE CONTAMINATED ENVIRONMENT

Reading	Maximum Time of Exposure in Contamination
Less than 0.2 rem/hr	8.5 hours per day
0.2 to 0.5 rem/hr	4.2 hours per day
0.5 to 2 rem/hr	No more than 1 hour per day
2 to 10 rem/hr	No more than 12 minutes per day
10-50 rem/hr	No more than 3 minutes per day or total of 21 minutes per week
50-100 rem/hr	No more than 1.5 minutes per day or 10.5 minutes per week
100 rem/hr	Do not go into contamination unless not doing so would mean your death

But even if you follow all these procedures, there's still a strong possibility that there will be contaminants in the air which you breathe or ingest. Therefore, let's look at some ways to reduce the damage they'll do to your body.

It should be remembered that the dangers posed by low concentrations of radioactive contamination in the air, or in food and water are rather insidious because the symptoms they create take so long to develop. Those who take absolutely no precautions will be in as good shape *initially* as those who take all precautions. Thus, if you are downwind from a major accident but see little immediate danger (since radiation is invisible), you might be lulled into inaction and run the risks of leukemia in a few years or cancer in 20 to 30 years.

This is exactly what can happen following a nuclear accident.

In such a case, authorities or politicians may try to minimize the danger, news reporters often don't tell precisely what the radiation levels are, and the average man-on-the-street who has no radiation detection equipment panics and does things that don't improve his chances of survival or finally just gives up on doing anything.

Thus, it will take 20 or 30 years to really see how bad the disaster is and — even then — it will be hard to tell whether a specific cancer death is the result of exposure to radiation or something that would have happened anyway. (This is why experts' predictions of premature deaths caused by the Chernobyl disaster varied from a few thousand to as high as 75,000. No one really knows until the actual statistics are collected.)

But... among those who take proper precautions, there will be a much lower death rate from cancers induced by radiation than among those who don't. While precautions won't guarantee that you or your family won't die of cancer 20 years down the line, your chances of not getting cancer will be a lot better than if you did nothing.

And wouldn't extra years of life be worth a lot and avoidance of a painful early death be worth the effort?

On the other hand, if you fail to do anything because of inadequate warning from officials or because you are unable to take all precautions (and assuming you only receive a mild exposure to radiation), you'll only face increased *risks* of leukemia or cancer. Such sickness will *not* be a foregone conclusion.

The risks of getting cancer or leukemia aren't as great as many people fear. If, for example, a population gets a low-level dose of radiation which doesn't cause short-term health risks, it may increase the percentage of leukemia cases in the population by as much as 30 to 40 times. That sounds pretty awful (and it certainly isn't anything to take lightly). But when the figures are actually examined, it isn't as bad as one might think.

Since the normal occurrence of leukemia is only 2 or 3 people per 100,000 people, that means that even if the rate increased by 40 times, only about one person per thousand would be affected. Thus, leukemia would still be very rare. The game appears to hold true for other forms of cancer as well.

What radioactive isotopes are produced in a nuclear accident?

The answer depends a lot on the type of accident. But, in general, the most dangerous isotopes would be the ones which mimic minerals that are normally needed by the human body. While the actual amounts of isotopes from a nuclear disaster would be small, the human body concentrates the isotopes in specific tissue or organs of the body. This means the concentration becomes high and the damage can become very severe over time.

Also, some organs of the body store and retain the materials for a long time. In such a case, a lot of damage can be done to the organ involved, especially if the isotope has a long half-life.

And what is a half-life?

As an isotope decays and gives off radiation, the atom turns into a more stable form and usually becomes less radioactive and therefore less dangerous. The time taken for enough of a radioactive isotope to turn to a non-radioactive form to cut the total radiation given off by the material in half is known as its "half-life."

It takes many half-lives before a radioactive substance loses all radioactivity; one half-life doesn't end its radioactivity. The amount only drops to half of the original. Thus, if a source of the isotope barium-140 were giving off 4 rems per hour, in 12.8 days (barium-140's half-life), the radiation reading would have gradually dropped to 2 rems per hour. After 25.6 days from the original reading, the level would have dropped to 1 rem per hour.

Half-lives of isotopes can vary from a few minutes to years.

Interestingly, radioactive isotopes which have long half-lives generally are less dangerous during any given exposure period as compared to those with short half-lives. This is because a short half-life means the substance is "decaying" faster and giving off radiation at a faster rate than its long half-lived counterpart.

Of the isotopes which might be involved in a nuclear accident, the most dangerous of the isotopes to human health are iodine-131, plutonium-239, cesium-137, barium-140; and strontium-90.

A number of iodine isotopes might be created by a nuclear accident. But most iodine isotopes have half-lives of less than a day, so they will be relatively "safe" by the time they reach a person or shortly after they have been "stored" in the human body.

One iodine isotope has a longer half-life, however. This is iodine-131; its half-life is 8.05 days. Because it is stored by the body in the relatively small thyroid gland, this gives it sufficient time to damage the thyroid gland before decaying.

Fortunately the body can tolerate, for short periods of time, much higher levels of iodine than it needs to maintain good health. This makes it possible to increase the intake of iodine and reduce the chances of the body storing the radioactive isotope of iodine and/or help the body to not retain the isotope if it has already been absorbed.

The most common way to get the massive dose of iodine is by taking a compound containing the chemical. The compound of choice is potassium iodide. From this solution, the body removes the iodine and — because of the large amount of iodine available — is more apt to reject the radioactive isotope. The potassium iodide mix most often recommended is Lugol's Solution.

Lugol's Solution is also known as SSKI. It's available in most drug stores and will keep for some time if it's stored in a dark brown glass bottle with a non-metallic lid.

Longer storage life of this chemical is possible if it is purchased in crystalline form from a drug store or a chemical supply house. To mix "Lugol's Solution" from crystals, a container is filled 60% full of the crystal, then water is added to the container so that it is filled to 90% of its volume. When the crystals are dissolved — except for a few remaining undissolved at the bottom — the solution is saturated and ready for use.

The recommended dosage of Lugol's Solution is 130 milligrams per day (about 4 drops) for adults. For children under one year old, 1/2 of this amount or 65 milligrams (about 2 drops) per day should be sufficient. The solution is to be taken from the first day that radioactive iodine may be in the area and continued (for up to 100 days) until the last of the material is detected in your area by government monitors.

This solution was given to 11 million children in Poland following the accident in the Chernobyl reactor number 4. One film clip of the procedure was broadcast a number of times on national TV in the US, and showed a school child taking the solution, crying, and making a terrible face.

There was a reason for that crying and face-making: potassium iodide tastes awful! When giving it to children (or "children of all ages," as the saying goes), try to take it with other water and food.

Potassium iodide tablets designed to be taken following a nuclear accident are also sold. These get around the problem with taste. Currently the US Government seems to be discouraging the sale of these; the thinking seems to be that giving the medication to those living around reactors when nothing is wrong might undermine the confidence of the public in the nuclear industry or that the tablets might be taken when they weren't actually needed.

These risks are small compared to the alternatives of not having them available if needed, however, and hopefully reason will prevail and they'll be readily available when needed.

In the meantime, these tablets are available commercially from a number of survival supply companies and might even be available at many local drug stores (or might be ordered through them).

If you plan to get such tablets, be sure to have the tablets stored away well ahead of time since there will undoubtedly be a panic run on such materials following a nuclear accident. (There was a small "run" on gas masks and potassium iodide tablets in the US following the Chernobyl accident half way around the world. Just imagine the panic if the disaster were closer to home.)

Since the dosage of each tablet may vary from source to source, be sure to follow the directions on the container of potassium iodide tablets.

Currently, there are experiments which suggest that a possible alternative to taking Lugol's Solution or tablets is as close as the family medicine cabinet.

New research seems to indicate that it's possible for the skin to absorb iodine so that it is carried through the bloodstream to the thyroid gland. While only lab animals have been tested at the time of this writing, the experiments indicate that a person should be able to get the iodine needed to counteract the radioactive iodine isotope by painting a patch of skin about the size of the person's hand with 2% tincture of iodine. While this remains experimental at this time, this last-ditch method might be employed if you don't have any potassium iodide tablets or solution available. (Do *not* take the 2% tincture internally; it's poisonous.)

One warning on the use of potassium iodide: while excessive amounts of iodine are not normally life threatening, the massive doses of iodine which the body will be getting may create bad side effects including hyperactivity or even the development of a goiter. So be sure you really are in danger before taking this solution.

Contact a doctor if any side effects are noted. Finally, anyone with an allergy to iodine should not use potassium iodide on their skin.

Plutonium-239 would probably not be present in a nuclear accident unless a nuclear weapon (or material to make one), a breeder reactor, or a terrorist attack were involved.

While a member of the nuclear industry once remarked that plutonium is so safe that he'd sit on a cube of it without any worry, plutonium is actually quite poisonous.

The reason the official might sit on a cube without ill effects is that your skin or clothing will shield you from some of the low-level radiation produced by plutonium (and possibly this official had in mind another isotope of plutonium, number 242, whose principle radiation is alpha particles rather than both alpha and gamma).

The catch is that if any dust particles containing plutonium are absorbed when it is inhaled into the lungs they will gradually damage the lung tissue. This damage finally will result in cancerous tumors years later.

Plutonium can also be absorbed into the bloodstream through the lungs to be deposited in the bones and liver. Again, it will damage these organs over time.

Because plutonium-239 has a half-life of 24,400 years and the body has a tendency to store it for many decades, it's quite dangerous. Any area which suffers an accident involving plutonium will probably have to be abandoned or undergo very extensive decontamination.

Unfortunately there is no chemical which can be used to displace plutonium in the human body. Avoidance is the only strategy. Avoid inhaling plutonium by staying out of contaminated areas or by filtering the air you breathe (more on masks later on).

Cesium-137 is not as much of a problem as radioactive iodine and plutonium because cesium-137 tends to spread throughout the body and is usually retained only a short time. Only a very large dose would be damaging to muscles, the spleen, and/or the liver.

Cesium mimics potassium in the human body; taking a potassium supplement would help reduce the intake of cesium.

Barium-140 and strontium-90 are two isotopes which mimic calcium in the human body. Strontium-90 is the more dangerous of the two since it has a long half-life of 26 years; barium-140 has a half-life of only 12.8 days. If either of these are ingested, the body stores them in bones and other tissue.

Fortunately, the human body will "choose" calcium over either if sufficient amounts of calcium are present. This makes it imperative to increase the intake of calcium if radioactive strontium or barium may be present.

The catch is, that many sources of calcium may become polluted with strontium-90 or barium-140. For example, cattle eating grass exposed to strontium contamination will concentrate the material in their milk; if you drink such milk, you'll increase your risks of ingesting and storing the two isotopes in your body rather than lowering your risks.

The winning strategy here is to buy a number of bottles of commercial calcium supplement tablets now and to store them away for emergency use. (If possible, purchase calcium from oyster shells rather than from bone meal; bone meal calcium may contain excessive amounts of lead.)

Uranium will be represented in the contamination produced by most nuclear accidents. Uranium is normally not retained by the body but large amounts can act as a chemical poison; such large doses damage the kidneys (because of its poisonous nature rather than its radioactivity).

If proper precautions are taken, by not eating contaminated food and by wearing a mask to remove any uranium dust in the air, a person should be able to avoid most of the dangers of uranium, except where the radiation levels are very high.

While low levels of radioactive contamination can pose problems from a long-term health standpoint, a few precautions and use of food supplements can help you avoid all this.

Next, we'll look at ways to avoid danger when high levels of radiation are involved.

TO FLEE OR NOT TO FLEE?

Whether to sit tight in some sort of shelter or flee for safety should be a fairly straightforward choice unless contamination is very low-level.

If it is low-level, you'll be able to take many of the precautions outlined elsewhere to minimize your exposure and reduce your chances of ingesting radioactive contaminants. Sitting tight at your home could sometimes be a good idea if levels of radiation were low.

Cities could quickly have their streets clogged during a panic; sitting tight might keep you from getting killed in a grid lock riot.

Likewise, looters might enter an area of low contamination. Depending on the degree of damage created by looters, insurance companies might decide you'd endured a riot. Riot damage isn't covered on many policies. Even if damage and theft caused by looters is covered, replacement of some valuables is hard or impossible if sentimental value is figured in.

And insurance companies may go broke trying to pay off the back-breaking bills run up during a nuclear accident. Many governments have also limited the insurance/claims/law suits that can be filed against the owners of a commercial nuclear reactor that goes haywire, too. Getting your money back gets to be an iffy proposition at best.

If you stay in your home to "guard" things, you can also watch to be sure any clean up work being carried out is done adequately or even direct or carry out some of the work yourself. Staying home is not without its advantages.

However, before you stay in any area downwind from a nuclear accident, you must be sure that radiation exposures will only be minor. The best way to judge the danger is to know how far away the source of the contamination is, which way winds are blowing, etc., and to have your own radiation detection equipment which can be used to measure any radiation which your area may be receiving.

Looters generally come from outside an area to help themselves to victims' belongings, though this might not be the case in a wide-scale nuclear accident. While you can't stop large-scale looting, you can minimize or even prevent any losses to your home (or business) with the proper tactics and equipment.

If you need to travel in an area that has been having trouble with looting, travel early in the morning. Looters prefer the darkness and will probably still be sleeping off their nightly activities if you travel in the early morning. When you're traveling, don't stop for any road blocks except those set up by the police or National Guard units. Stopping at a roadblock set up by the lawless could get you killed. Be sure to carry some good ID papers so that you can prove to National Guard or law-enforcement officials that you belong in the area.

When it comes to protecting your home from looters, you should *not* keep a low profile. Rather you will need to give a "show of strength."

Looters tend to steal only from safe areas. They also often fear authorities or those who appear to be in power. Therefore, a show of strength is anything you can do which makes it appear that you're stronger than the group of looters.

A good example of this occurred during the Miami riots several years ago. In the middle of the burnt-out section of Miami, where rioters and looters destroyed a large number of buildings, was a trailer court. The mobile homes remained untouched during the mayhem. The home-owners sat at the entrance of the trailer court with guns: handguns, hunting rifles, shotguns, .22's...anything that looked dangerous and showed they meant business. Not a shot was fired; their show of force prevented the looting and destruction of their homes.

For an effective show of force, several things are required.

Good lighting is important. Lights show that you are home and not afraid of letting anyone know that you're there. Lights also mean that the looters can be seen and later identified by you.

Lights can be a problem if the power goes off. One way around this is to use an emergency generator or lights run off your automobile. In such a case, lights can be especially effective as they create the illusion of an island of civilization in a sea of trouble.

If your family joins with other members of your neighborhood to protect your area, be sure that some sort of identification is used to keep you from mistaking any of the neighbors for looters or vice versa. "Uniforms" can be arm bands, yellow shirts, anything that will become a badge that quickly identifies friends from intruders and which can be placed over protective clothing or suits.

Uniforms can attract trouble or repel it depending on how you use them. Be sure that the uniforms don't look like those of the police or National Guard; looking like either of these groups may cause you to be attacked by sniper fire by any "whackos" wandering about.

Use uniforms only in the area you're protecting. If you need to travel, change to regular clothing so that you won't be seen as a target by looters who might take out their frustrations on someone they view as an authority.

Fire is the looter's tool so you should be prepared to fight fires. Fire extinguishers should be where they can be used quickly. Chances are good that city water supplies will be shut down during a nuclear emergency. But you might have a hose connected to an outside faucet just in case (this would also be of use later in decontamining your house).

Firearms should only be displayed when looters get too rowdy or dangerous. A weapon should be displayed *only* if you are prepared to use it. If you do use it, expect to receive return fire and act accordingly. Don't fire warning shots; such shots will probably be interpreted as shots fired at the looters. *Never* fire indiscriminately into a crowd; aim for looters who are actually endangering you.

Never let a group of looters get close to you. They may have concealed weapons and if they charge, your time to shoot to defend yourself will be very short. If you have to shoot in such a situation, aim for leaders, those who are armed, and those at the front of the group so that they will block the path of those behind them. Once the fight starts you'll be fighting for your life.

The sooner you know of an emergency, the sooner you can prepare for and react to the danger. One good way of doing this is to use a radio capable of picking up national emergency/weather stations; such radios are helpful in alerting you to an emergency. These radios are readily available at most electrical stores. (Radio Shack is probably one of the better sources of inexpensive equipment.) Most ideal is an automatic radio which turns itself on to alert its owners when the national weather station sends out an emergency signal at the beginning of a civil defense broadcast. Such a signal may give you advanced warning and is ideal for alerting you if an emergency takes place in the middle of the night.

(One excellent radio for this purpose is the "Weatheradio Alert" currently sold by Radio Shack for $40. The unit can pick up National Weather Service and CD warnings from any station within 40 miles. It has a backup 9-volt battery so the unit will continue to operate if the AC power goes down, and it sounds an automatic alert tone to warn you of a special warning for your area.)

Once you have a warning that an emergency is taking place, monitoring police and other emergency radio bands (inexpensive scanner radios to do this are also available in most electronic stores) will also help you discover exactly what is happening rather than waiting to get a muddled, late, and even censored report from the media.

Once you discover the source of radioactive danger, check the direction of the wind and try to determine where air-borne contamination will be headed, what direction the evacuation will be headed in, etc. Later, you may be able to receive emergency instructions over the weather station as officials start to handle the problem.

If you determine that a major accident has occurred and that contamination may be major, your strategy will be simple. Because the radioactive material used in industry and nuclear reactors has a very long half-life, the solution is to *get out immediately.* And if you're in a major urban area then you should get out *very* quickly.

If you chose to stay in a heavily contaminated area, you'd practically have to hole up in some type of shelter for years or even decades because of the time it would take for most nuclear materials to decay to safe levels. (Again, this is *not* the case with fallout from nuclear weapons; this material has a short half-life so that staying in place until the radiation level sinks to a safe point is a viable strategy.)

In most countries, the government will undoubtedly make an effort to help evacuees and perhaps even furnish clothing, food and shelter. (The amount will depend on the country, the area contaminated, and the money available for such efforts.) Even so, it would be a good idea to have a relative, friend, vacation home, or some other "refuge" to which to flee. Accommodations for survivors of a nuclear accident would initially be quite crude at best and law-and-order might be nonexistent for hours or days as refugee camps form following a disaster.

During the mass exodus from a contaminated area, things could be extremely volatile as well, especially if panic spreads through the people in the area or a traffic jam snarls up the mass exodus. Too, law enforcement manpower will be stretched to the limit and almost nonexistent during the evacuation.

This is not just speculation.

In the aftermath of many large-scale disasters in the recent past, there have been stories of people taking vehicles at gunpoint, stealing belongings from survivors of accidents, or even raping and murdering refugees. In short, during a nuclear accident, it could be critical that *you* be able to defend your family if you are to survive.

Unfortunately, technology doesn't offer any "humane" way to stop a criminal wanting to do bodily harm to you. Stun guns, tear gas, etc., are all notoriously ineffective when used in actual encounters; firearms are the only weapons which have enough range to keep an attacker at bay, allow the weak or injured to protect themselves, and are capable of defending the user against multiple attackers. (Interestingly, recent civil defense studies in the US show that a firearm is one of *the* things that citizens would take with them in the event of a major evacuation of a city. That was what they said even when instructed not to do so and when only filling out a questionnaire where they might have fibbed!) All in all, it would seem prudent to carry a firearm *provided* you know how to operate it properly (firearms in the hands of the untrained are dangerous to the criminal and innocent alike).

While having a firearm means you should be prepared to use it to defend yourself, the appearance of a firearm often has a deterrent effect. While this deterrent is not often recognized by those opposed to the ownership of firearms by citizens, in fact many crimes never occur because the potential victim produces a gun and is prepared to defend himself with it. Recent studies suggest that at least as many occurrences of non-shooting, self-defense use of firearms happen as do shoot outs where the owner of a firearm protects himself from a criminal. Therefore, you are actually *less* apt to be hurt (or to hurt someone) when you are prepared to defend yourself with a firearm.

While almost any firearm can be used for self defense, some are much better than others. Because of the probability of dense areas of people where the weapons might be needed, a rifle would not be ideal; centerfire rifle cartridges fire bullets capable of penetrating the human body. Rifle bullets might conceivably injure innocent people even if direct hits were scored against antagonists. Therefore, the best bet would be a pistol; a shotgun or carbine chambered for a pistol round are possible second choices for those who do not wish to own a pistol (or are currently prohibited from ownership by law).

While a firearm should not be carried if the owner isn't prepared to use the weapon, just the appearance of a gun in the hands of a determined-looking person can be a great deterrent to attack. How many looters would be quick to challenge this man in a contaminated area? (Shown is the Colt AR-15 rifle.)

There are currently four calibers of pistols which are relatively easy to control during recoil while still having the power to take a criminal out of the fight if one of the bullets strikes a vital area. At the same time, these pistol chamberings aren't overly powerful, so the risk of harming innocent bystanders is greatly reduced (although care still must be exercised).

The four best pistol chamberings for self defense are the .38 Special, 9mm Luger (also known as the 9mm Parabellum), .357 Magnum, and the .45 ACP. The .38 Special is marginal but will work in its "+P" loading which is readily available commercially. The other three have sufficient power to do the job, with the .357 Magnum tending to have a shade too much power.

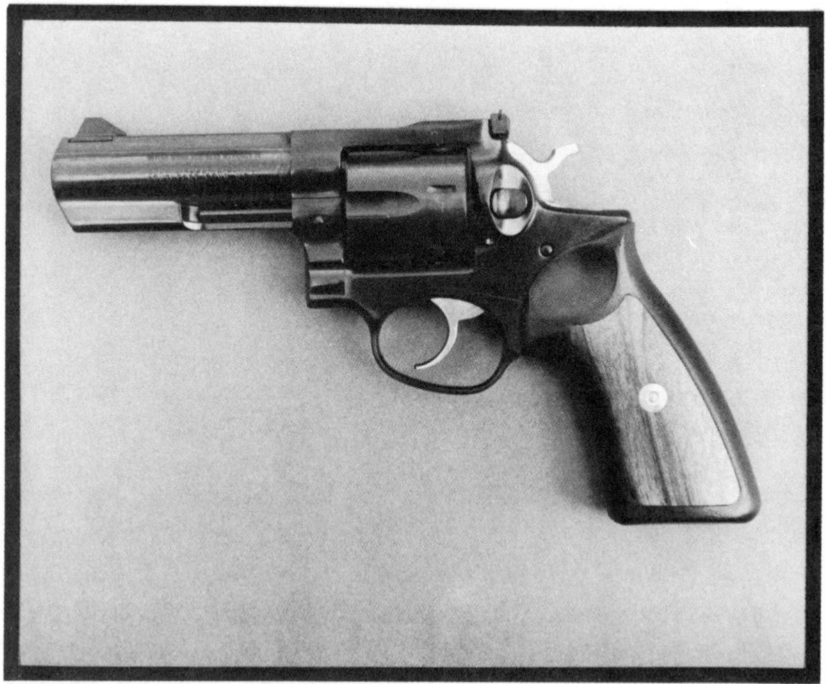

Sturm Ruger's GP-100 is one of the author's first choices of .357 Magnum revolvers.

With any of the four, the ammunition used must have hollow-point bullets capable of expanding; don't use FMJ (Full Metal Jacketed) bullets as they tend to travel through the human body doing only a minimum of damage while being capable of wounding anyone behind the criminal you're shooting at.

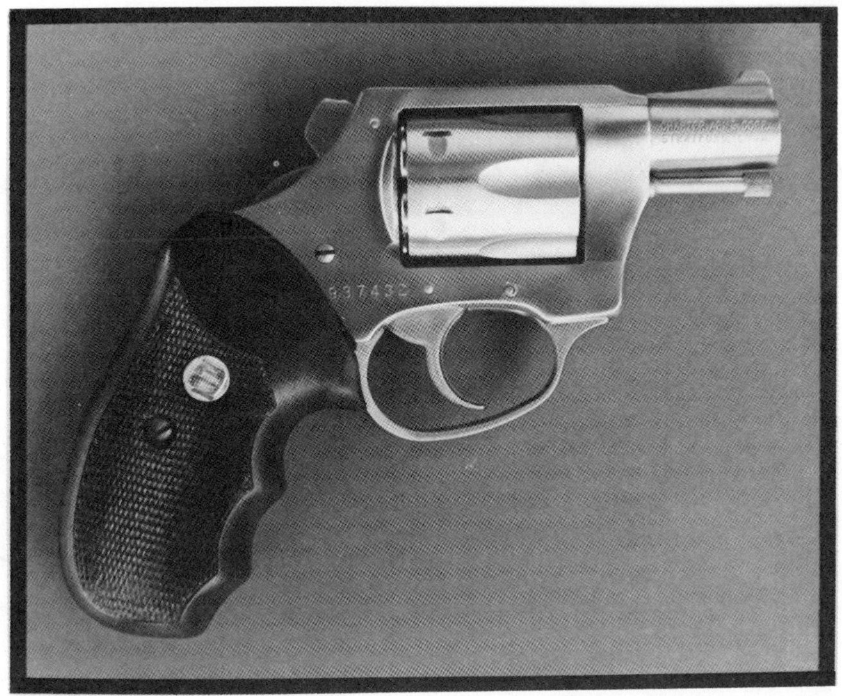

Charter Arms offers a number of excellent "hide away" revolvers. Shown here is their stainless Undercover model with Charter Arms neoprene grips.

The .38 Special and .357 Magnum are normally used in revolvers. While the Dan Wesson, Colt, Smith & Wesson, and other well-known companies offer fine guns, currently the first choice in these calibers are the revolvers marketed by Charter Arms and Sturm, Ruger and Company. These two companies offer well-made firearms that are reasonably priced. Ideally you'll purchase a revolver chambered for .357 Magnum; it can chamber either the .357 or .38 Special rounds giving you the choice of firing the more powerful .357 or the milder-recoiling .38 Special.

In the .357 chambering, the Charter Arms Target Bulldog and Ruger GP-100 would be first choices. If you wish to only fire the .38 Special and want a short-barreled "hide-away" pistol, the Charter Arms Undercover, Off-Duty, or Police Undercover models would all be ideal.

For those wanting a .45 ACP automatic, Colt's pistols are among the finest. Shown here is the "chopped" officers' model currently being offered by Colt Firearms.

The Arminex Tri-Fire is an excellent pistol which is actually an updated version of the Colt .45 Auto. The Tri-Fire can also be modified with kits to fire 9mm Luger or .38 Super.

Despite what you may have heard, the .45 ACP cartridge doesn't create much worse wounds than the .38 Special loaded to +P standards. Because of the limited magazine capacity of the .45 automatic pistol — 7 rounds — it also doesn't offer nearly as much combat capability as modern 9mm automatic pistols. (For a detailed discussion of how the .45 got its unwarranted notoriety, as well as how to create or choose combat ammunition, see my book *Combat Ammunition,* which is available from Paladin Press for $19.95. You may also be interested in my *Combat*

Automatic Pistols which is also available from Paladin.)

If, however, you decide you wish to have a .45 ACP pistol, probably the first choice is the Colt .45 Auto with the Arminex Tri-Fire also being a good choice. Other copies of the pistol which may suit your purposes are the Springfield Armory 1911-A1 copy of the pistol or Smith & Wesson's double-action 645 (which has an 8-round magazine).

The author's first choice among automatic pistols is the Beretta 92-F. The pistol is super-reliable, well made, and should be in production for a long time.

First choice in the automatics are the new double-action 9mm pistols. These pistols offer a lot: internal safeties which make it possible to carry the pistol without the need of having a manual safety engaged; double-action first shot — no cocking the hammer — just draw and fire; and large capacity magazine giving the user from 15 to 18 shots before reloading (such a capacity might be a necessity if you're facing a number of attackers). Of the 9mm automatic pistols, the best is the Beretta 92-F (recently chosen as the side arm for the US Military); the 92-F pistol is well made and promises to be in production for a long time.

Other good choices in the 9mm pistols are the Steyr GB, Sig-Sauer P-226, and Smith & Wesson 659 or 669. (At the time of this writing, the Ruger P-85 9mm pistol was unavailable for testing. It looks like an excellent pistol, however, and should merit consideration if it is available to you.)

An excellent book on the "how to" of using a pistol in combat is *The Complete Book Of Combat Handgunning* by Chuck Taylor (available from Paladin Press). While the book is a bit behind the times in how it rates .45 ACP over other chamberings, the "how-to" and tactics parts of the book are well thought out.

You should remember that reading a book on shooting won't make you an expert pistolero. You need to go out and practice and become used to shooting your weapon. Shooting also is ideal to get your pistol to perform properly; most guns reach their peak of reliability only after having 100 or 200 rounds of ammunition fired through them.

Another category of handguns which should be mentioned are the "assault pistols." These are actually just semi-auto pistols but are modeled on submachine guns. Because of their dangerous looks, they could offer some extra deterrent from attack if you were carrying one. However, you can *not* depend on looks to defend you. As mentioned before, if you carry a weapon, you must be prepared to use it. Assault pistols are small enough to be concealed under a coat or jacket and can be easily carried in a bag or suitcase. They have very large magazine capacities — up to 36 rounds or more for those chambered in 9mm (which seems to be the best round for this type of pistol).

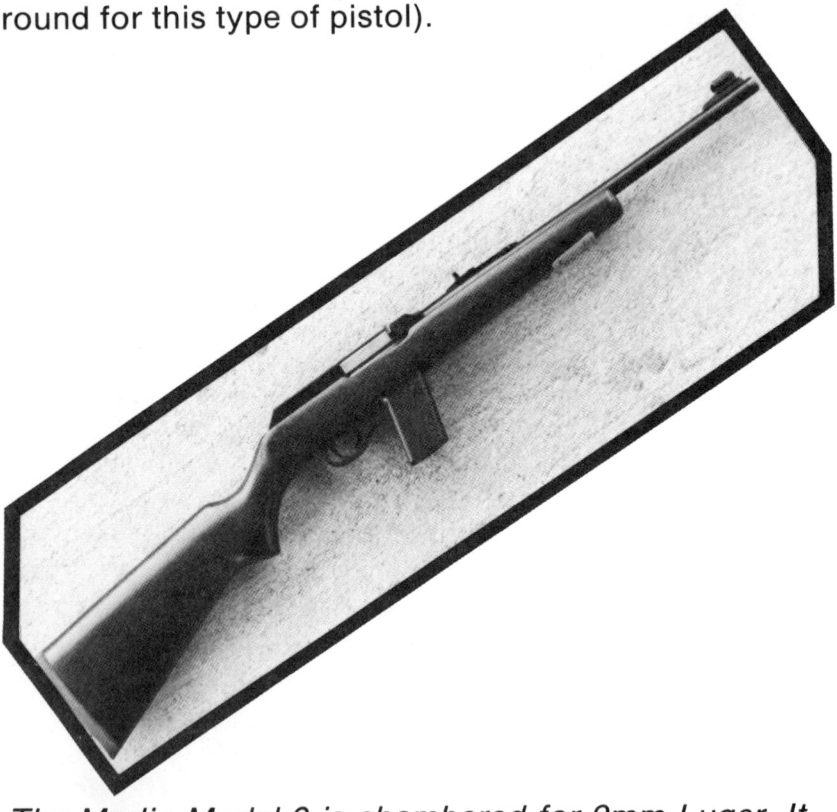

The Marlin Model 9 is chambered for 9mm Luger. It is a good choice for those who don't wish to purchase a pistol and want to avoid the overpenetration of most rifle bullets.

Among the best assault pistols are the various models of the Goncz High-Tech Pistols and the

Intratec TEC-9. (The Goncz pistols have a slight edge over the TEC-9 since the TEC-9 requires a second hand to operate the safety and is not as well balanced.) Uzi pistols are overly heavy and way too expensive, though they are good; both the TEC-9 and Goncz pistols have price tags that make them less expensive than regular pistols to purchase.

Pistol-caliber carbines might also be considered by those living in areas which have banned handguns but not long guns. The light-weight Goncz carbine (which is nearly identical to the Goncz pistol except for the longer barrel and stock) is leader of the pack here with the Marlin Model 9 in 9mm or Model 45 in .45 ACP being good — and inexpensive — second choices.

The lever-action Marlin 1984CS is ideal for those who want a good self-defense rifle that looks rather innocent in an antique sort of way.

If you would like to have the extra power of the .357 Magnum, the lever-action Marlin Model 1984CS Carbine is ideal; the rifle looks like an antique and won't attract undue attention — but it is capable of shooting quickly and effectively. (Looking non-military might be an important consideration in times of martial law or emergency. Having a "cowboy" rifle rather than a military-looking weapon might save your firearm from being confiscated at a National Guard roadblock.)

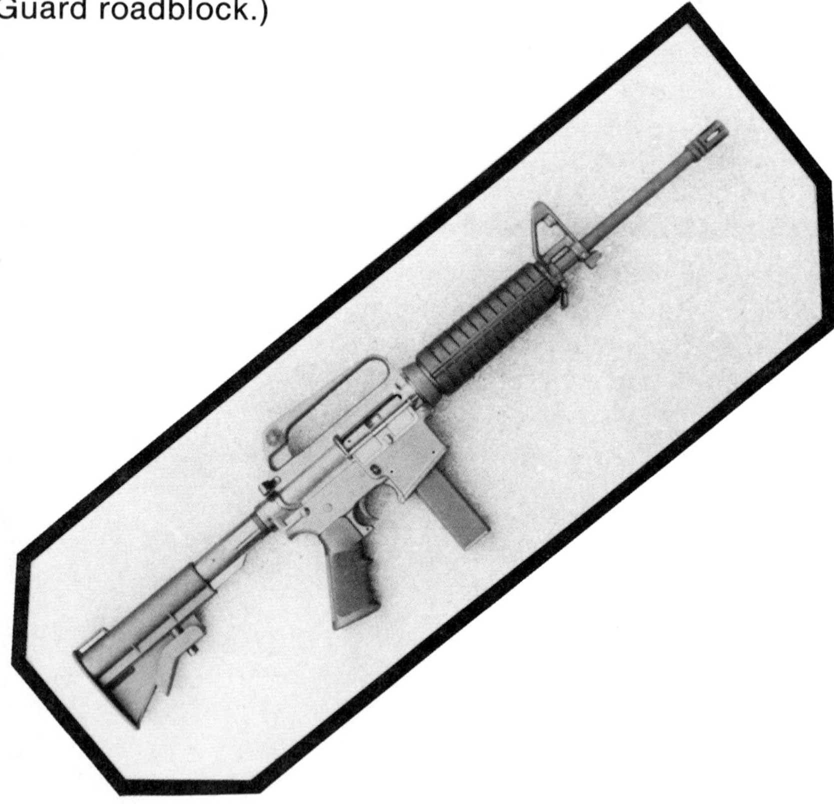

The Colt AR-15 Carbine chambered for 9mm Luger has the look of a full-fledged assault rifle while keeping the low penetration of pistol-caliber bullets (with suitable ammunition).

More expensive — but menacing looking — are the Colt AR-15 9mm and the Heckler and Koch HK-94 carbines. These carbines both fire 9mm Luger but look like full-fledged assault rifles; they also command higher prices.

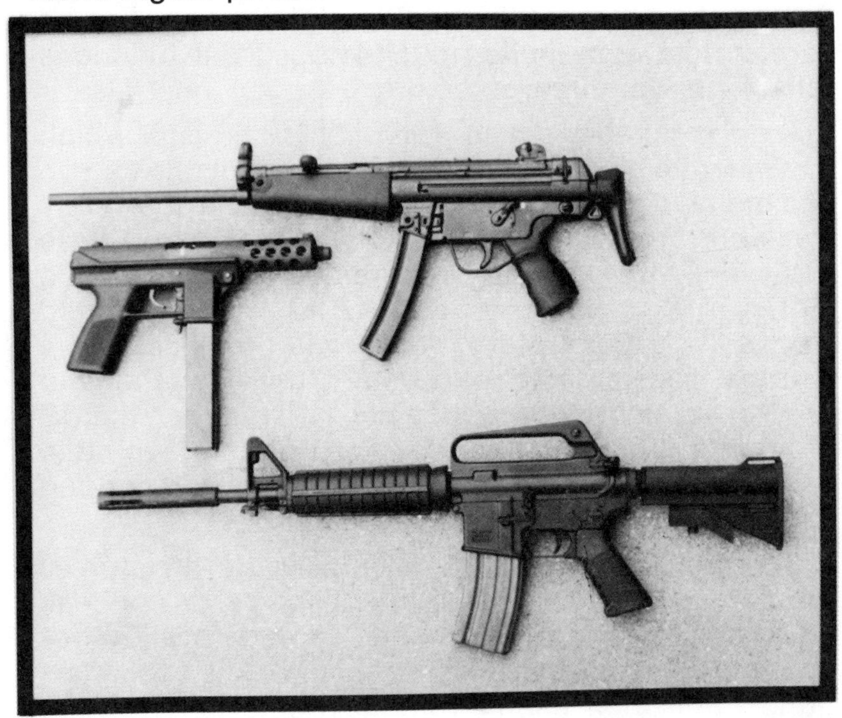

Three possible solutions to the self-defense needs of the victims of a nuclear accident: (top) the HK-94 which fires 9mm Luger; (center left) the TEC-9 which also uses 9mm Luger; and the SGW AR-15 carbine which fires the powerful .223 Remington cartridge.

Whether you're using a pistol or a carbine, the best choice in ammunition for the four chamberings listed above are Winchester Silvertip hollow-points, CCI's Jacketed hollow-points, CCI's Lawman, and Federal's hollow-points. (CCI's Blazer ammunition is also good in handguns but the aluminum cartridges should *not* be used in the blow-back assault pistols or carbines.) *Always* use hollow-point bullets for self defense; FMJ bullets are dangerous to innocent bystanders and fail to give the maximum stopping power which a round is capable of delivering. And be sure to test ammunition out to be sure it functions reliably in *your* firearm.

A good holster is also essential for a pistol which may need to be carried yet be readily available for use at a moment's notice. Probably the best buy currently available are "Uncle Mike's" nylon holsters. These cost very little and are available in almost all gun stores. The nylon and lining of the holsters mold themself to your pistol for a custom fit and the material is strong and rot resistant, unlike leather and other materials. If you want a flap holster, then the US Military's UM-84 ambidextrous holster (available from Brigade Quartermasters for $40) is a very good choice.

Shotguns give a lot of "overkill" and are not as precise as pistols nor as easy to carry or conceal. Shotguns do not give enough spread to allow for much error in aiming (despite being called "scatter-guns") and their useful combat range is pretty well limited to 30 yards even with buckshot or 100 yards if slugs are used. And only slug loads have a lot of recoil, small magazine capacity, and are awkward to use. All in all, the shotgun has a lot going against it.

However, the shotgun does almost always stop anyone who suffers a direct hit by the numerous pellets in one of its loads. And a shotgun is very intimidating when you're looking down its barrel.

Remington's 870 and 1100, Mossburg's 500 Security, Winchester's Defender, and Ithaca Model 37 DSPS are all good. Semiautos with gas actions absorb recoil but are slightly less reliable in adverse environments than are slide action ("pump") shotguns. The Benelli M121, Browning, and Franchi shotguns are also good semiauto actions which allow quick follow-up shots; their recoil, however, is nearly as great as that of the slide-action shotguns.

If possible, get a shotgun with a magazine extension tube to increase the firepower of the weapon; some models of shotguns come with the extension tube already in place which can be a savings in time and money. Most gunsmiths can purchase any of the extension tubes for the shotguns listed above. For do-it-yourselfers, Choate Machine and Tool offers extension tubes and other accessories for most of the shotguns listed above. Installation of the extension tube is simple; you just remove the end cap on the magazine tube and replace the magazine spring with the new one in the Choate kit, and screw the extension tube on the shotgun. Your shotgun is ready to go and will supply a number of extra shots.

Choate and others also offer a number of shotgun accessories. Folding stocks, flashlight mounts, sling swivels, etc., might all be of use for some owners of "combat" shotguns. Just be sure you really need any accessories you may wish to add to your shotgun.

Barrel length of a combat shotgun should be limited to 26 inches with an 18- or 20-inch barrel being best. "Deer" barrels with rifle-style sights are also available in 20- or 22-inch barrel lengths for most shotguns; these are ideal for those used to using sights for aiming.

The best chamberings for combat with shotguns are 12 and 20 gauges; 20 gauge offers a little less "oomph" in most loads but gives less recoil. When possible, use birdshot rather than slugs or buck shot to avoid overpenetration in urban areas. Even birdshot is capable of penetrating several walls of a house so don't fire indiscriminately with a shotgun. Number 4 buck shot is ideal in areas where overpenetration or hitting innocent bystanders isn't a concern. Again, the best companies for reliable ammunition are Winchester and Federal.

Because shotguns are sporting weapons for the most part and have a very short range, it is often possible to locate trap ranges where you can practice with your weapon and gain skill in using it. Shotguns, despite the terrible wounds they can create and their greater danger in short range combat, are generally viewed by the non-shooting public as "more acceptable" weapons as compared to handguns or "assault rifles." Therefore, it's often possible to own a "hunting weapon" in areas which have all but banned the ownership of handguns.

An excellent book which looks at various techniques for using the shotgun is Chuck Taylor's *The Combat Shotgun And Submachine Gun* (available from Paladin Press). Practice reloading your shotgun and then pray you won't have to in combat; reloading is tedious and easily muffed in combat. When reloading, be sure to push the shell well into the magazine of the weapon. Failure to push it all the way in may cause it to pop back into the action of the weapon, jamming some brands of shotguns.

If you're expecting a real major social upheaval or have remained with your house in low-level contamination, then you *might* need an assault rifle to defend yourself. The best of these is the Colt AR-15 and its variants. Of the various models, the AR-15 A2 Sporter is first choice since it is capable of using a wide range of ammunition and its fast barrel twist creates maximum-size wounds with all types of bullets. The Ruger Mini-14 is also a good rifle and, though not as rugged, quite reliable and usually only half as expensive as the AR-15.

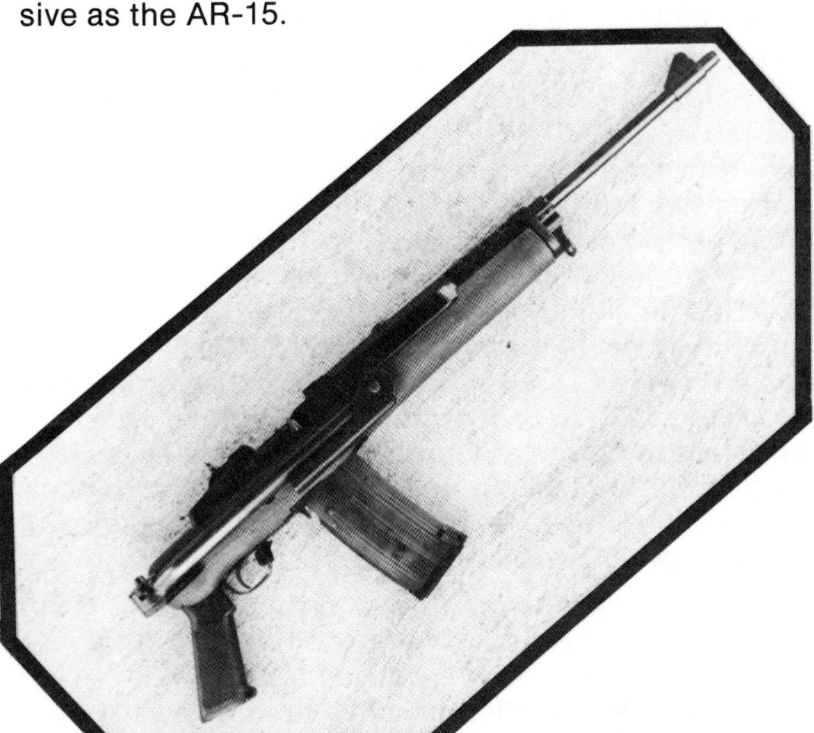

For those wanting a rifle, the Ruger Mini-14 is a good choice. (Shown here is the ranger model with folding stock. The extended magazine is marketed by Ram-Line.)

If you purchase an assault rifle, get four or five 30-round magazines and good ammunition for it; both Federal and Winchester sell good ammunition topped with Spitzer-style bullets (with an exposed lead tip) which are best when dealing with human targets. Though the bullets fired in a .223 are small, they create massive wounds and can easily penetrate a house. You must be careful where you fire them as they can travel several miles and still be capable of killing an innocent bystander. Again, practice is essential for surviving combat where such a weapon is called for.

How much ammunition? If you're in an urban area or in an area near a large city, you could need a lot if things really get bad. There is no rule of thumb here, but many feel that the user of an assault rifle should have at least a thousand rounds to fall back on in an extreme emergency. With a pistol or shotgun, the numbers of rounds you had in reserve could be much smaller; because of the short range of both, your chances of surviving more than a few "close encounters" is smaller. Chances are, you could get by with just four or five boxes of shotgun or pistol ammunition during any break down of law-and-order which might occur during a major nuclear accident. (This is, of course, assuming that law-and-order will again re-establish itself in your area.)

Remember to never fire at an angle which will cause bullets to go upward — the bullets from a rifle can travel miles and still be lethal. Likewise, never shoot into the air. When possible fire downward from a higher position when using a rifle. Pistol bullets also travel quite far though not as far as rifle bullets. Shotgun pellets lose much of their power after about 100 yards; by several hundred yards they will be relatively safe. It is also possible to fire a shotgun into the pavement ahead of looters and hit them with ricocheting birdshot pellets; the pellets will sting but usually won't penetrate the skin. The catch to this tactic is that it is easy to aim too high and catch people in the feet. Too, even if successful bounce shots are achieved, the crowd will assume that you're shooting to kill. Whatever weapon you're using, don't shoot unless there is no other way to protect yourself.

If you carry a firearm, don't get cocky or overconfident. Many an armed man has been brought low by a man with a knife or even just a club or a broken bottle. Don't take extra chances because you are armed. Be as cautious as you would be if you were unarmed.

Of course, firearms won't magically get you "home free."

During a mass evacuation, communications will be overloaded, many areas may be heavily contaminated (or rumors to the effect may be spread among evacuees), and many people will probably panic.

If the civil defense/government hasn't gotten its act together or made the order to evacuate, the citizens may start evacuating in a disorderly fashion themselves (this has been what happened historically in times when great dangers were perceived by the citizen populations in many US and European cities). Even if the CD gives orders for an orderly evacuation, if you're in an area where everyone owns a car, you can expect a disorderly evacuation at best. (If you need an idea of what it will be like, check out the highway near a football stadium right after the game.)

Heavy traffic and/or a panic evacuation means you'll have a hard time getting out of the area if there is a large population. The only way to get around this problem is to have all you need to take with you ready to go. That means you should be prepared to travel light. You'll not need a lot of supplies in the area you go to (especially if you have a charge card or two and some cash squirreled away for such an emergency); the government or relief groups will probably take care of your needs even if you have little more than a shirt on your back.

But first you must get out of the heavily contaminated area.

A bug-out bag allows you to make a quick dash away from an area of heavy nuclear contamination. A number of excellent bags suitable for such use are available on the marketplace. Shown here is Brigade Quartermasters' excellent B4 flight bag. (Photo courtesy of Brigade Quartermasters.)

Ideally, you'll have everything you need packed in a duffle bag or lightweight back pack (or several bags, one for each member of your household). Keep the bag (or bags) ready to go in the trunk of your car or in a closet at your house. Such "bug-out" bags allow you to simply grab the bags and a firearm or two, collect the children and spouse, get into the car, and drive off, if a major emergency calls for you to leave the area. (A number of good, rugged bags and back packs suited for use as bug-out bags are available from Brigade Quartermaster; Uncle Mike's also has a number of small pouches and belt packs which would be ideal for carrying many of the odds and ends from a bug-out bag.)

If possible, each bug-out bag should have a strap or shoulder harness so that you can carry it easily with your hands free should you be forced to leave your car and travel on foot. Ideally the bag would also be made of a muted brown or green fabric so that it won't stand out should you need to hide. (While you'll probably have no need to hide, if you do, having a glowing orange bag could spell your death during a social upheaval.)

What should go into the bug-out bag?

You need food, water, and shelter to survive.

But chances are you'll be able to walk to safety in a few day's time at the worst. So you could get by without food if you had to. If you don't want to rough it, get several packets of freeze-dried food and some hard candies (Sierra Supply is a good source for freeze-dried food as well as a lot of other military & surplus gear mentioned in this chapter.) Both store for a long time and are relatively lightweight for the energy they can supply.

Water can probably be gathered along the way unless you're in a very dry area. But you'll need to be able to carry it; get one or more military surplus canteens and water purification tablets (more on purifying water later). If your bug-out bags are stored in an area where they won't be exposed to below-freezing temperatures, then water could be stored in canteens.

Since there's no way to predict *when* a nuclear accident might occur, you should be prepared for the worst of weather in your area. If it gets extremely cold in your area, the simple solution is to grab winter clothing as you leave your home; don't try to store too much in each of your bug-out bags. Just don't forget to get the heavy clothing even if it is unseasonably warm when you evacuate the area. You may have to spend the night outside or in a poorly heated area. (A couple of ponchos do make sense in a bug-out bag, however, in case a rain storm greets your trip and you have to go on foot.)

Other essentials for *each* bug-out bag would include a *small* first aid kit (with a large gauze pad for dealing with a major injury), a good pocket knife, a *small* compass, some nylon cord, a small flashlight, a candle, matches (or a butane lighter), a large plastic bag, a small (hotel-size) bar of soap, one box of ammunition for any of your firearms, sunglasses, and a good pair of *broken in* hiking boots and heavy socks.

You'll probably want to add one of the Guillory Aegis I suits to each family member's bag and perhaps a radiation detector and portable radio in one of your family's bags. Anyone needing special medication or foods should also be supplied, perhaps with a little extra in case supplies are short at the refugee center in which you may find yourself.

Finally, be sure that your bag contains records of your bank accounts, contract agreements, social security numbers, insurance policies, deeds, etc., etc. As mentioned before you should also have an emergency fund of cash and a credit card if possible. If you have cash you may be able to buy your way out of trouble; a bank card would be useful once you reached an area which was safe to stay in. (Many people don't realize that it's possible to own several bank credit cards. For example, if you already have a VISA card with X State Bank, you can also get another with Y National Bank. All you have to do is go into the second bank, get a form, and fill it out. Having a card set aside for emergency use would be a good investment of a little time and a small yearly service fee.)

Some authorities recommend jewels, jewelry, or silver/gold coins for use in such an emergency (as well as others). Such items are *not* ideal commodities in a crisis. The average man on the street has no idea what they're worth and will probably not give you anywhere near a fair trade for any such items. Too, having such things in your possession might make it appear that you'd looted a coin or jewelry store on your way out; looting may not be treated too lightly during a major crisis.

If trouble seems to be occurring at a near-by nuclear plant or other facility, you'll need to be careful you don't get sucked into unwise decisions or suggestions from others. Don't listen to "gossip" whether it is bad news or good news. When it looks like you should "head for the hills," take off. Don't wait.

Waiting for a crowd to gather on the highways will only guarantee that you'll get snarled in a good example of grid lock. You have no duty to die in a traffic jam just because others wait until the last minute to protect themselves.

It is also hard to predict just how much help or "neighborliness" will be taking place in the aftermath of a disaster. While most people cooperate and help each other following a major disaster, they may not always do so, especially in a very large disaster where great numbers of people are suddenly killed. If you are part of a major nuclear disaster, you may discover that things deteriorate to a very dangerous level very quickly.

For example, the post-disaster euphoria that is usually experienced by survivors of a disaster was *not* experienced by the people who survived the Nagasaki and Hiroshima nuclear blasts at the end of WWII. In fact, the people in the two populations didn't even cooperate to carry out their own rescue work as is normally the case in disasters.

Social scientists who study such things have found that the more sudden and unexpected a disaster and the greater its devastation, the less apt people involved in the disaster are to help each other.

A nuclear accident would probably not be too widespread. But it could be. And it would be unexpected. So the idea that "we'll all just pitch in together and work things out" may not be realistic when thinking of major nuclear accidents.

Major accidents also create rather bizarre behavior among many people.

Those who have been having minor mental problems often go "off the deep end" and even normal people may not deal with problems rationally. Relief workers have found that following a tornado or other major and sudden devastation, people can often be found standing in the rain raking the lawn or trying to care for their damaged flower garden while their houses have been opened like a can of worms and all their valuable possessions are blowing away or being soaked by the rain.

Likewise, people may try to continue at their jobs as if nothing had happened, ignore warnings to evacuate an area, or try to go to work in areas totally destroyed by a disaster.

Unfortunately, social pressures — especially in a democratic culture — tend to look at individuals as having a voice in what should be done — even if they are totally off their rockers. Thus, you may experience "social pressures" to act as if nothing is wrong or follow others in their illogical behavior. Doing so could be fatal during the first critical hours following a nuclear accident. Resist pressure from those who don't — or won't — grasp what is happening and ignore advice that "everything is going to be all right." Such advice could cost you and your family your lives.

When you've arrived at a safe area, you may also discover that you or family members feel guilty for not dying when many others have perished. This is a normal feeling following a disaster and should soon pass. Try to remind anyone feeling this way that they have no real reason to feel guilty. If extreme feelings of depression or stress show up, try to support the family member affected and don't be hesitant in seeking professional help from relief counselors.

Doctors, food, and other supplies as well as law-and-order (in the form of the National Guard) should arrive at the scene of a nuclear accident hours or days following the event. Remind others of this and "hang on" until help arrives.

In order to survive, you must take care of yourself and your family. While it would be ideal to help others if possible, don't worry about situations you can't control or help with. Dwelling on problems which aren't yours will only lead to dangerous depression.

Be prepared to help and rebuild things and ignore problems you can't do anything about.

With a little preparation, a few supplies, and a dash of luck, you *can* survive a nuclear accident and evacuate to a safe area quickly and with a minimum of danger.

DETECTION EQUIPMENT

Unfortunately, nearly all the radiation detection equipment which is currently available on the surplus market is designed for nuclear war use. While such equipment is capable of detecting many of the radioactive materials from a nuclear accident, the equipment is adjusted for very high levels of radiation. This means that you might be getting low levels of exposure while your meter didn't show that any was actually present.

Another problem with equipment designed for nuclear war use is that it often is not capable of picking up beta or alpha radiation.

While equipment detecting only gamma radiation is fairly safe to use for detecting most nuclear contaminants, a few isotopes which can affect human health don't give off gamma radiation.

(Of the most dangerous isotopes — to human health — which might be involved in a nuclear accident, iodine-131 and barium-140 do give off gamma — and beta — radiation and plutonium-239 gives off gamma — and alpha — radiation. But cesium-137 and strontium-90 give off *only* beta radiation.)

Having a meter that only detects gamma radiation might have serious consequences in a nuclear accident; your meter might show a safe level of radiation when you were actually facing hazardous exposure levels. This could be especially dangerous when checking food for contamination.

(Since none of the dangerous particles listed above give off only alpha radiation, you could get by with equipment that didn't detect alpha particles but would warn of beta and gamma radiation.)

While beta radiation not detected by such gamma-reading-only meters would be stopped by clothing or even your skin, receiving exposures to the radiation over any period of time (or ingesting it with your food) could produce very serious health problems.

Most surplus civil defense equipment for sale to the public in the US is also very old. Since the US has greatly reduced the money spent in protecting its civilian population from nuclear war (though Congress and the military continue to spend large sums on equipment and shelters for themselves), most of the surplus meters found for sale need special batteries that are a bit hard to find, and worst of all, are not re-calibrated despite the aging of the components in the units.

All this means that you'll not be finding much in the way of useful equipment in the used/surplus area.

It is possible to purchase a number of new, modern meters designed for nuclear war/civil defense use. Unfortunately these are calibrated to read in higher 0-200 or 0-500 rems ranges. This means that the low end of their scales is not accurate as compared to a meter which gives 0-500 millirems readings. On the high range scale, the needle may be showing less than 1 rem but you have no idea how much less.

One meter that overcomes a little bit of this problem (and which is my favorite "nuclear war survival" radiation meter) is the Plessey PDRM 82, which is manufactured in England and imported into the US by Guillory and Associates.

This "hi-tech" meter uses 3 standard "C" cells (which can give it 400 hours of continuous use) and has a microprocessor which actually checks the unit when you turn it on to be sure that it is working (the word "TEST" appears during the test; "FAIL" shows if the unit is malfunctioning; "BATT" appears if the batteries need to be replaced). The PDRM 82 continues to test itself while in use with a built-in beta source which it compares to the radiation in the environment.

Best of all, the PDRM 82 has a digital liquid crystal display so that it is nearly impossible to misread the display and, if the 300-rem level is exceeded, "300" flashes on and off on the display. The unit has a specified shelf live (without batteries) of at least 20 years when stored in a dry area; you can buy it and store it away for later use without worry.

The catch with the PDRM 82 is that its digital reading is from 0.1cGray/hr — 300 cGray/hr (remembering that a CentiGray is equal to a rem). This makes the level of reading too high for use with most nuclear accidents (but ideal for nuclear war use). But *if* a similar unit should be marketed for low-level use which detected both gamma and beta radiation, this would be just the ticket for nuclear accident use.

It is probable that other companies will soon follow the lead set by the PDRM 82. (For those interested in purchasing a PDRM 82, cost is $375 and the unit is currently being imported into the US exclusively by Guillory and Associates.)

One way around the problem is to just not purchase detection equipment. Then your tactic to deal with the possibility of undetected radioactive contamination is to take precautions as if the area you're in is "hot" with contaminants. But it would be much safer to have a way to measure radiation in order to avoid dangerous areas as well as for conducting decontamination procedures.

A better solution is to obtain a radiation meter designed for the nuclear industry and use that in a nuclear accident emergency.

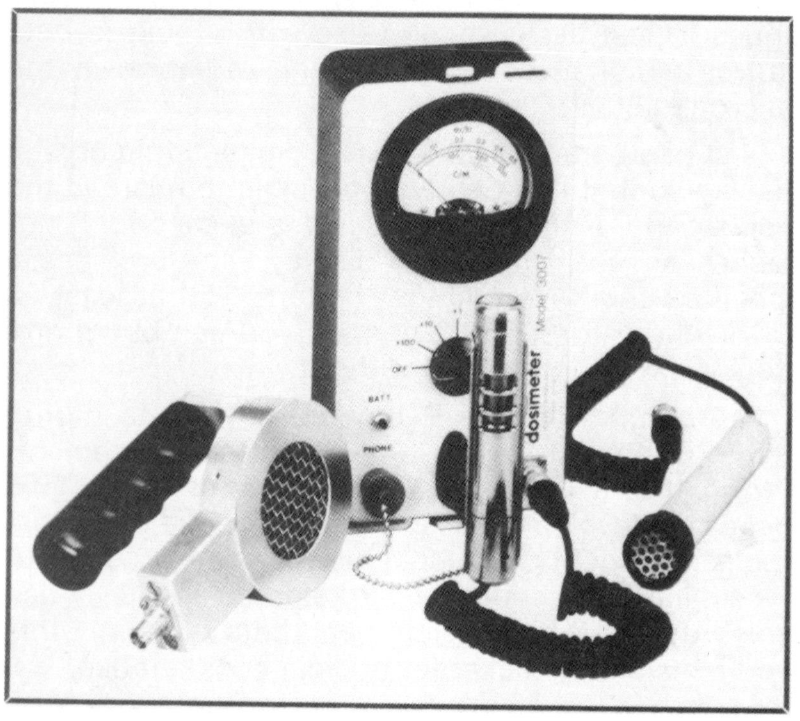

Dosimeter Corporation's 3007 Survey meter with the 3011 probe is ideal for detecting contamination generated by a nuclear accident. (Photo courtesy of Dosimeter Corporation.)

Many of these industrial meters are capable of detecting both gamma rays *and* alpha/beta rays, too. While many such meters are expensive, they will allow you to know whether or not an area or object is dangerously contaminated.

One excellent industrial radiation meter is currently being manufactured by Dosimeter Corporation. This "3007 Survey Meter" has three low-level ranges of 0-0.5 millirems/hr; 0-5 millirems/hr; and 0-50 millrems/hr. By adding Dosimeter Corporation's 3011 probe to the 3007 meter, it is possible to detect alpha, beta, and gamma radiation.

Dosimeter Corporation's 3700 meter has three ranges and detects both beta and gamma radiation. (Photo courtesy of Dosimeter Corporation.)

The cost of the 3007 meter is high: $290 for the meter, $150 for the probe. The meter uses 2 "D" cells for power. The only catch is that the range of the meter is low. If you were in an area of *very* high radiation, the meter might "lock up" or give inaccurate readings; however, the 3007 is designed to not jam at readings as high as 1 rem/hr so this is unlikely with this meter. (While you should evacuate an area with such high readings, it is wise to be aware of "lock up" problems which most low-level meters have. If you suddenly start getting low readings, you may be in a hot spot.)

Another similar Dosimeter Corporation meter worth considering is the 3700 which costs $350 with a gamma/beta probe. This meter has three ranges: 0-0.5 millirems/hr, 0-5 millirems/hr, and 0-50 millirems/hr. The unit's only shortcoming is that it doesn't detect alpha radiation, but chances of being in an area contaminated with material giving off only alpha radiation is rather remote. This unit will also quit giving accurate readings or even lock up when levels of radiation go beyond 1 rem/hr. A "window" on the probe must be opened on this unit to detect beta radiation; the unit runs on common "D" batteries.

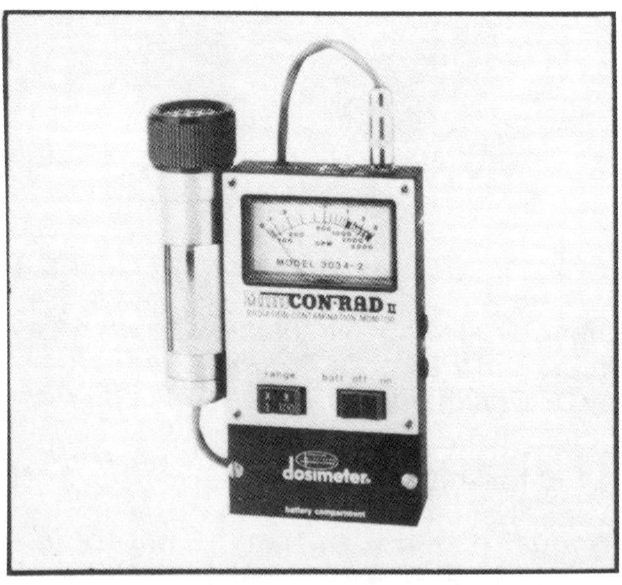

Dosimeter Corporations' Mini-Con-Rad II detects alpha, beta, and gamma radiation when used with its external probe. The unit is light-weight and easy-to-use. (Photo courtesy of Dosimeter Corporation.)

Also worth considering is Dosimeter Corporation's Mini-Con-Rad II; this pocket-sized meter detects only gamma radiation with a built-in probe or, with an external probe, will detect alpha, beta, and gamma radiation in the range of 0-500 millirems/hr. The unit has an internal speaker to give an audible alert, as well as visual with the unit's gauge. The Mini-Con-Rad meter uses a 9-volt transistor radio battery and costs $550.

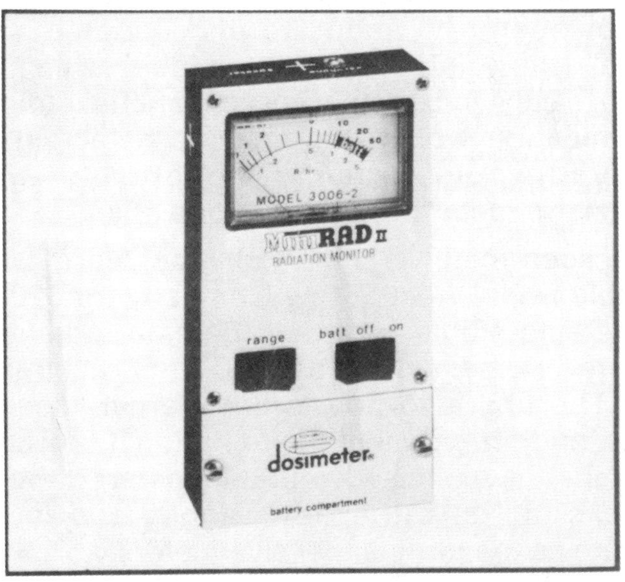

Dosimeter Corporation's Mini Rad II fits into a shirt pocket and runs on a nine-volt battery. The unit has a built-in speaker in addition to its front gauge. (Photo courtesy of Dosimeter Corporation.)

Dosimeter Corporation makes another pocket-sized meter which comes close to bridging the gap between nuclear war and nuclear accident usage. This is the Mini Rad II. Like the Mini-Con-Rad II, the Mini Rad II is pocket-sized and works on a 9-volt battery.

However, the Mini Rad II measures only gamma and X-ray radiation. The trade-off is that it has two ranges of detection, one in millirems per hour and the second in rems per hour. If a meter is needed to serve double purpose, the 3036-2 model of the Mini Rad II is probably the best bet. It has a 0-500 millirems/hr and 0-50 rems/hr range and costs $325.

If all these prices seem a little steep, there are budget-priced meters. But the less expensive meters are not nearly as accurate and usually lack the "probe" (the wand-like detector tube) which allows you to easily locate the source of the radioactive contamination which the meter is detecting. However, you would certainly be better off with an inexpensive meter rather than none at all.

One such meter is the "Radiation Alert" which is available from Direct Safety Company for $200. This unit detects alpha, beta, and gamma radiation and has three ranges which cover 0 to 50 millirems per hour. The unit also boasts an internal "beeper" to give an audible signal of the radiation it is recording. Care must be exercised with this meter, however, as it "locks up" at rather low ranges and is not as accurate as the more expensive meters.

Direct Safety also offers the RDX-Radiation Monitor. This unit detects only beta and gamma radiation. It uses a 9-volt battery and has a switch which activates the meter only when it is held down (this keeps you from accidentally leaving the meter on). An internal speaker gives an audible click when radiation is detected and the gauge gives one continuous reading from 0 to 10 millirems per hour. One excellent feature with this meter is that it gives an audible warning if its scale has been exceeded and the meter is "locking up." Cost is $110.

Another similar unit is the Monitor 4 Radiation Alert pocket detector (available for $165 from Direct Safety). It is capable of detecting alpha, beta, and gamma radiation and operates over three ranges to encompass 0 to 50 millirems/hr. The battery-operated detector will fit into a large pocket and has an audible beeper as well as a built-in meter.

Regardless of which type of meter you have, remember that there is background radiation in all natural environments. Getting an occasional reading means your equipment is working, not that you're in danger. Only when you start getting abnormally high readings do you have something to worry about.

Most meters have a strap with them; if not, it is easy to improvise a strap from a piece of string, an old belt, or similar material. A strap can be used to carry the meter over the shoulder; it's also useful for holding the meter close to the ground while checking an area for contamination.

(When walking in a contaminated environment, when the meter is at chest or waist level, your feet will be getting a higher exposure than the meter reads and your head, lower. The reading of the meter will give a rough average of your total exposure rate. Holding the meter on a strap will give the reading near the ground but not necessarily the rate of exposure you are receiving, which will be slightly lower.)

Meters only give the rate of exposure. Since total exposure is what determines the damage to your body, it would be best to have an instrument which shows your total exposure, too. Dosimeters are devices for determining your total exposure to radiation since the device was last zeroed.

Modern dosimeters are about the size of an ink pen and are electrostatically-charged, sealed tubes that lose their electrical charges with exposure to radiation. By looking through the eyepiece on one end of a dosimeter, it's possible to read the amount of radiation it's been exposed to. By wearing a dosimeter, a person can have a good idea of how much radiation he's been exposed to.

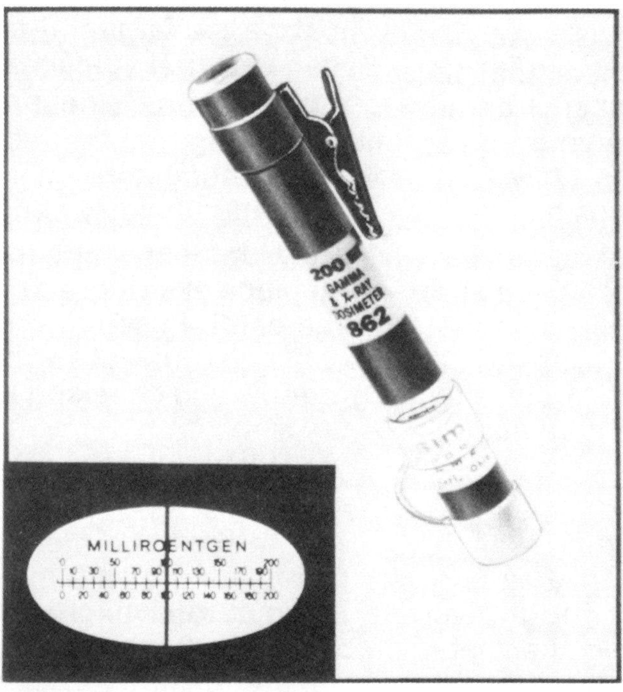

Modern dosimeters, like this sold by the Dosimeter Corporation, are about the size of an ink pen. Inset shows the picture obtained when viewing the scale through one end of the tube. (Photo courtesy of Dosimeter Corporation.)

While in the theoretical world it might be possible to get by with just a radiation meter, in the real world where readings vary greatly from place to place, a dosimeter would be very useful for finding the total radiation exposure you've received.

Dosimeters need chargers to give them their electrostatic charges. As radiation goes through this charge, it is reduced which, in turn, changes the reading of the dosimeter. Thus, the amount of loss of electrostatic charge will show the amount of radiation a dosimeter has been exposed to. Since the electrostatic charge remains constant only for a month, the dosimeter must be recharged at least 12 times a year to work. That means that you'll also have to purchase a charger if you buy one or more dosimeters.

A personal dosimeter can be worn on a chain around the neck, clipped to a belt, or carried in a pocket. If you have only one dosimeter for a group of people, it is possible to place it in a central location and get a rough idea of everyone's exposure (though individual dosimeters are *much better.*)

Personal dosimeters should be worn so that they give a good idea of what your average body exposure is to radiation (if contamination is on the ground, your feet will be receiving a higher dose and your head a smaller dose).

Don't place a dosimeter on an extremity such as in your boot or on a hat. This could give erroneous readings since the dosimeter would be closer or farther away from the source radiation.

Finally, dosimeters detect *only* gamma and X-ray radiation. You should be aware of this fact in case you are exposed to some "off beat" contamination that consists only of alpha and beta radiation.

A wide range of scales are available with dosimeters.

Those used by medical and industrial workers generally measure millirem, milliRoentgens, milliSieverts or milliGrays; those designed for emergency civil defense purposes often measure in whole rem (or equivalents).

Since your major concern if trapped in a contaminated area for any time will be your risk of short-term radiation sickness, a high-scale meter measuring rems rather than millirems would probably be the better choice. Otherwise, you might discover that your low-range dosimeter had gone over the top of its scale; in such a case you'd have no idea whether you had been exposed to only a few rems or a major dose which dictates your finding immediate medical help.

While a high-range dosimeter won't give you an overly accurate idea of how much exposure you've received if you are in an area of low-level contamination, it will give you a warning if you're nearing the dangerous levels of more than 100 or 200 rems. So, a dosimeter that has a scale of 0-600 or 0-200 rems would probably be best for non-medical/non-industrial users.

(A high scale dosimeter would also be useful if your area received fallout from a nuclear weapon — whether from a terrorist attack, nuclear accident, or an all-out war. In such a case, having a high-range dosimeter would allow you to deal both with a serious accident or radioactive fallout from a nuclear weapon.)

In heavy contamination or during times of long exposure when the dosimeter would have to be recharged, it would also be important to keep careful records of dosimeter readings to determine total exposures. Knowing that you could spend an extra 15 minutes decontaminating an area without becoming ill would be very useful information.

A carefully kept log is the way to keep track of total exposures; at regular intervals, take a reading from your dosimeter and log it; then recharge and zero the dosimeter(s) so that you can continue to get readings. Add up the readings to figure out total doses. Your log need not be more than an old notebook; just be sure to put down the date, your name (if there is more than one person's dosimeter reading being logged), and the reading from the dosimeter.

(A carefully kept log may also be useful in predicting possible long-term health risks following a nuclear accident. Knowing that added risks may be present would enable a person to be aware of possible symptoms and — with proper medical help — improve survival chances if cancer or other diseases developed years later.)

Dosimeters are read by pointing one end toward a light source (sunlight, electric light, or even a candle) and looking through the viewing end of the dosimeter so that you can see the scale inside the dosimeter. A hairline indicator shows the total dose the dosimeter has received since it was zeroed.

Some dosimeter chargers may also have a small light built into them which allows dosimeter readings to be taken. Great care must be taken to keep either the charger or dosimeter shielded so that the dosimeter isn't recharging accidentally. Accidental recharging of the dosimeter will cause the reading to be lost.

Whenever you read a dosimeter, point it at the same angle each time. Readings can change by several rems if the unit is pointed horizontally and then vertically, etc. You should also check the zero of the dosimeter after it has been charged from the same angle that you'll be taking the readings from. Otherwise, a "zeroed" dosimeter may give a reading that is high or low at the angle you'll take readings from.

Store dosimeters in a charged state and recharge them every month or two. Compare readings of the various dosimeters you have before recharging to see if they are all getting the same readings from background radiation and that one isn't malfunctioning.

Before using dosimeters which have been stored in an uncharged condition, charge them, allow them to stand for half an hour, and then recharge and rezero them. This will allow the dosimeters to give precise readings.

The best source of both dosimeters and chargers is Dosimeter Corporation. Their 686 dosimeter is ideal, with a reading from 0-600 rems; another good dosimeter, the 638, is similar to the 686 but reads in the 0-200 rems range. Each has a clip on it so that it can be held in a pocket or clipped to clothing and a plastic cap to protect the dosimeter from accidental recharging or damage to the contacts. The 686 is being phased out by the company; cost for it is $60. The 638 carries a $100 price tag.

Dosimeter Corporation's 909 Dosimeter Charger is well designed and uses a D-cell to work (which is ideal since AC power could be off during a nuclear accident). The charger has a bulb (and a spare bulb inside) so that you can take readings from the charger (just remember to leave the plastic caps on the dosimeters). The unit will charge and zero other types of dosimeters as well as those manufactured by Dosimeter Corporation. Cost for the charger is $90.

Meters and dosimeters can be sealed in plastic ziplock bags to simplify decontaminating them if you're in a heavily contaminated area; just take the bag off and discard it.

If you live near a radioactive waste dump or a nuclear plant, then there are also automatic radiation alarms which would give you an automatic warning if any radiation leaks should occur.

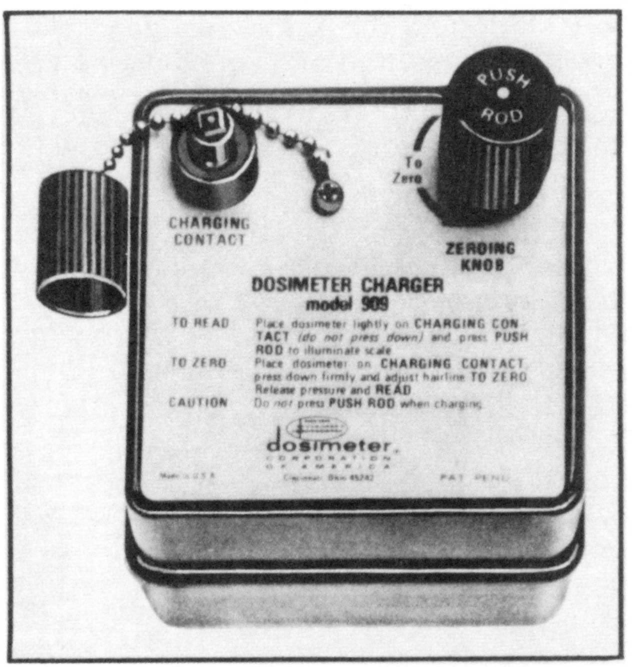

Dosimeter Corporation's 909 charger is well designed and is battery operated, making it ideal for emergency use. (Photo courtesy of Dosimeter Corporation.)

One such unit is the Micronta Radiation Monitor/ Alarm (available at some Radio Shack stores for $40). The unit is similar to a smoke alarm and runs on a 9-volt battery. When the Micronta alarm is exposed to gamma rays or X-ray radiation of 4 millirems/hr or more, the alarm goes off. The unit comes with a chart that allows you to calculate your total exposure to radiation by figuring the time the alarm first goes off.

The Micronta unit does *not* detect alpha or beta radiation so you should be aware of whether or not such radiation may be given off by any radioactive sources near you.

Dosimeter Corporation has a pocket-sized detector of gamma radiation which "chirps" when radiation is detected by the unit. Called the RAW-2, the unit has an internal unit to set the rate of chirping into one of three ranges. Cost of the unit is $175 (which seems a little steep considering the fact that the unit doesn't have a meter to show the actual range of radiation to which it is being exposed.)

The RAW-2 "chirps" to warn the user when certain levels of radiation have been exceeded. (Photo courtesy of Dosimeter Corporation.)

Dosimeter Corporation also makes the SuperDAD which gives both a reading of total exposure and "chirps" with radiation exposure to sound an alarm when a certain exposure has been reached. This unit would be ideal for working at decontamination work but could be less than ideal for detecting/warning of hazardous contamination. The unit costs $375, making it less than ideal for most non-industrial buyers.

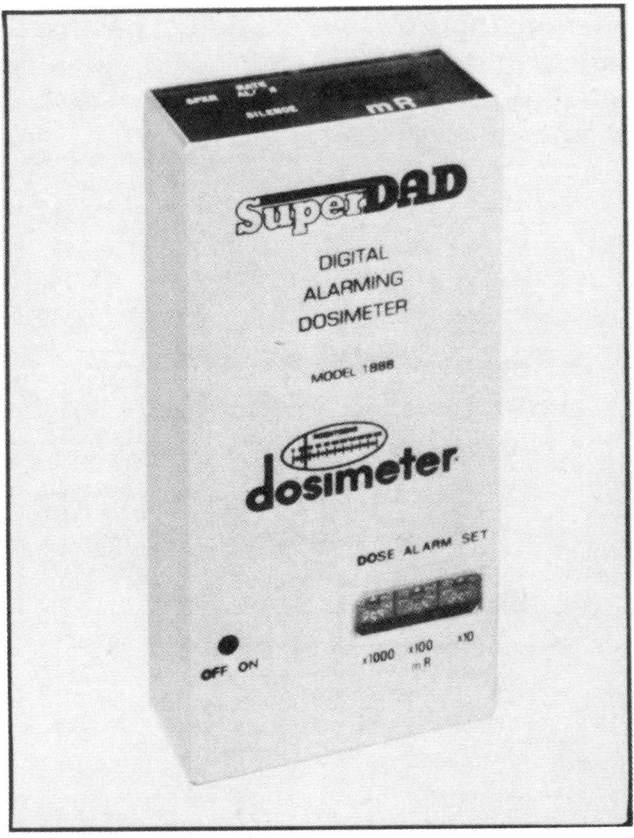

The SuperDAD is a digital alarm which "chirps" when exposures are becoming high. It has a four-digit readout on its top. (Photo courtesy of Dosimeter Corporation.)

Dosimeter has taken yet another approach to the alarm problem with their Area Alarm Monitors. These units sound an alarm when certain levels of radiation are exceeded and also have a gauge which allows the user to see exactly what levels of radiation are being recorded. The 3096-2 is the low-range, more sensitive version of the meter with a scale moving from 0.1 to 2000 millirems/hr. The 3090-2 covers 1 millirems/hr up to 100 rems/hr. Both meters detect gamma and X-ray radiation and cost $650 each.

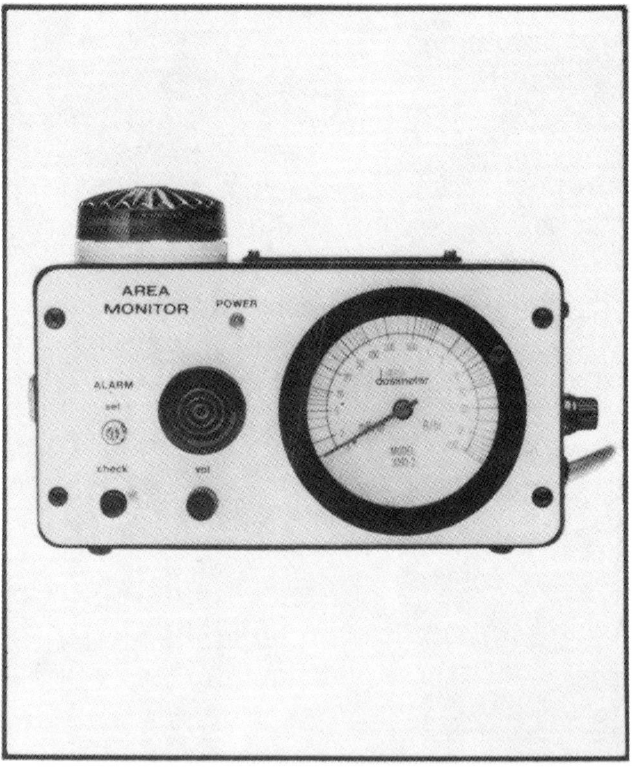

Dosimeter Corporation's 3090-2 is an area alarm monitor which gives both visual and audio warnings of how much radiation is being encountered by the user. (Photo courtesy of Dosimeter Corporation.)

Whatever equipment you're using, it's important that you not confuse whether you're taking readings of total radiation or the rate of radiation exposure. Total dosage and dosage per hour (which is what most radiation meters give) can be quite different. A reading of 120-rems per hour could mean a total exposure of only 2 if you were in the area for 1 minute. That would hardly be worth worrying about by emergency standards.

On the other hand, if you stayed in an area giving a 120 rems-per-hour reading for 5 hours, the total exposure would be 600 rems. You'd be very sick and would probably die if you didn't get medical help.

Total exposure is the most important consideration for your safety. A dosimeter is the easiest way to determine this. At the same time, a radiation meter is essential to show when you're in a dangerous area where your total exposure might become excessive.

So ideally you'll have both a dosimeter for each person in your family and at least one meter in order to stay safe. If you have only one or neither, you should not attempt to stay in any area which is contaminated, and carrying out decontamination procedures will be much harder to do and it will be impossible to check to see if you've missed any small bits of contamination.

Last, but certainly not least, don't cut things too close when getting radiation readings.

When you're nearing the limits of exposure to radiation that you can risk without harm, wrap things up and get to an uncontaminated area. This is especially important since nearly all dosimeters and meters have rather high error potentials with various types of radioactive sources. Even quality dosimeters, for example, will give readings of plus or minus 10 percent of actual levels. That means that your 100-

rem exposure could be as high as 110 rems. Over time, this exceeded level could take a great toll on the physical level.

Always play it safe, when you can do so.

PROTECTIVE EQUIPMENT

Even if you don't have a radiation meter or dosimeter, you can still avoid becoming contaminated by taking a few precautions.

First is avoiding dusty areas which may be contaminated by nuclear dust from an accident. If you must move about in such an area, then the second precaution is to wear clothing which is more or less dust proof and which can be discarded when you reach an area which is uncontaminated.

It's important to realize that such a protective suit will *not* give protection from gamma radiation which will be found in most types of contamination. The "protective suit" will give protection from beta and alpha radiation only; in effect it only makes it possible to keep contamination away from your skin so that you will have less trouble getting cleaned up when you reach a "clean" area.

A protective outfit can simply be heavy clothing which you've sealed up around the wrists, ankles, button holes, etc. A quick and dirty way of sealing up an improvised suit is with duct tape (which is available from most hardware stores); shoe strings or other cords can also be used to quickly seal up the sleeves and legs of an outfit. Better than improvised suits, however, are the "high tech" suits designed for use by the nuclear industry. These are usually made of Tyvek which is a disposable plastic non-woven material which allows air to move through it but filters out dust.

Whatever type of suit you have, it's important to also furnish covering for your feet, head, and hands. Otherwise, you'll manage to get dust particles in your hair, under fingernails, or in your shoes and socks. Covering these areas can again be improvised, an old pair of rubber rain boots or plastic garbage bags taped in place on your feet, a plastic-bag hat, and rubber dishwasher gloves. But "high tech" coverings are also available which are much better; surgical gloves and the Tyvek suits which come with hoods and feet in them are superior to almost anything you might improvise.

To protect your face, goggles are ideal. Since eyelids can trap radioactive dust in them and some contaminants may even be absorbed by the eye, keeping your eyes protected is important. An air filter or dust mask of some type is also essential if you're to avoid breathing in radioactive dust.

Most modern gas masks have filters available for them which are designed to protect the wearer from nuclear particles (by filtering them out) and will also absorb many radioactive gases. Therefore, such a gas mask is ideal for moving in a nuclear contaminated area. But if you don't have a mask, even just a handkerchief or a towel wrapped over your mouth and nose (the more layers, the better) will improve your chances of not inhaling a speck of radioactive dust and will give good — if not perfect — protection.

Of course, the easiest and quickest-to-use outfits are the suit "kits" which are composed of equipment designed for the nuclear industry. Such a suit, complete with mask and goggles, is relatively inexpensive. A number of suit kits are available. Probably the best currently available is the Aegis I suit from Guillory and Associates.

The Aegis I "kit" contains a Tyvek, hooded coveralls with feet in them to cover your shoes; a dust/gas respirator; goggles (which can be worn over most eyeglasses); and rubber gloves. The suit goes over your clothes and shoes so that it's quick to don if you find yourself in a contaminated area. The reusable mask accepts standard size industrial filters which can be replaced and which will remove a number of dangerous chemicals as well as many radioactive gases and dust particles.

For families with small children, Guillory also offers the Aegis I suits in small sizes so that children can also have a suit for protection (although a little duct tape can quickly alter an adult-size suit to child proportions). Medium and large sizes are available for adults. The Aegis I suit is small enough to be kept in a car trunk or your bug-out bag; and costs $49.50.

A similar suit — the Redi-Safe — is also offered by the Direct Safety Company for $16. Unfortunately the kit doesn't contain as good a mask as the Aegis I; the mask is only a 3M disposable filter which will not absorb radioactive gases. The mask will stop heavy dust, however, and its price is certainly competitive. The kits come only in medium sizes.

The only catch with either of the suits is that there is some gap between the goggles and the filter. Therefore, for maximum protection, a Tyvek suit should be coupled with a good gas mask.

If you decide to purchase a gas mask rather than a complete suit kit like those above, it's possible to purchase the Tyvek (or a similar fabric "Kaycel") suits separately.

Hooded Tyvek overalls with built-in boots are available from Direct Safety Company for $6.95 a suit. The white coveralls are cut full to go over clothing and are available in small, medium, large, and extra large sizes (product number N-05-314).

Military boots and gloves designed for use with NBC (Nuclear/Biological/Chemical) equipment are sometimes available on the surplus market. These are generally expensive and hard to find, however, so a better bet is to simply purchase rain boots and rubber gloves.

While suits with built-in feet (like the Tyvek overalls listed above) don't require boots, adding boots makes it less likely that you'll rip the material when walking over rocks or other sharp objects. In such a case, rubber boots could give you some extra protection from accidental contamination.

Pull-over rain or snow boots offer a better potential seal than do zipper- or clasp-closure boots. Duct tape should be used to seal up the pants-leg-to-boot-top area, as it is with other openings on an improvised suit.

Improvised foot coverings can also be created with small plastic garbage bags but these are generally only good for very short periods since they tend to sag and are easily punctured. You must use duct tape with these to fasten them on tightly.

Rubber gloves should be lightweight to allow easy movement and control when you need to run equipment. Dishwashing gloves are ideal provided you won't be doing any type of work which might puncture and rip the gloves; small punctures wouldn't be too serious; major tears would. Cloth gloves over rubber gloves may be used when doing light work but the cloth gloves should be disposed of as soon as possible since they could become full of radioactive dust.

If you need to work, better protection is offered by heavier industrial gloves. These can generally be found in hardware stores. As with other protective clothing, be sure to use duct tape to tightly seal them to your suit's sleeves.

If you have to improvise a suit from street clothes, heavy clothing, or layers will give more protection than lightweight clothing, and its effectiveness can be increased by limiting the air flow through it. Tuck everything in, button everything up, and use duct tape to seal up areas where the clothing overlaps (the waist band, shirt front, shirt sleeve openings, etc.). Woven fabric would need to be discarded or decontaminated after it was worn once, so it would be easy to quickly go through a lot of clothing during a period of decontamination work or the like. Non-woven rain gear, ponchos, etc., or closely woven fabric like that of a nylon wind-breaker would be easier to decontaminate though it would be uncomfortable in warm weather.

Coupled with a good suit, a modern gas mask can give you a lot of protection from radioactive dust particles. With a proper filter it can remove many radioactive gases which might be inhaled and absorbed into the body through the lungs. Many masks also give protection against a number of chemical accident materials so that the mask can serve to give protection from a wide range of potential industrial chemical accidents as well as nuclear accidents.

There are a lot of military surplus masks as well as industrial masks on the market which are suitable for use in nuclear-polluted environments.

The "working part" of a gas mask is its filter; if you have a bad filter or even the wrong one in your mask, it may not give you much or complete protection. Filters remove dangerous radioactive gases by chemical bonding between chemicals in the filter and the gas moving through it. They also remove dust particles — and radioactive dust — by physically filtering particles out of the air.

Both the bonding and the filtering cause filters to have a limited life after which they may fail. It also means that the heavier the concentration of contaminants and dust, the more quickly the filter will become saturated and clogged, unable to remove any other contaminants from the air moving through the filter.

When this happens, modern filters will limit the air coming through them. This is good news and bad news. The good news is that you can tell when the filter is starting to fail, the bad news is that the mask may start leaking when the filter starts to clog up so air flow is decreased through it. (Because of the finite life of a filter, it's wise to spend as little time in a contaminated area as possible since it's hard to predict how high the concentration your filters are working against or when they will start to fail.)

Because the greatest immediate danger to you comes in the form of small dust particles which you might inhale, the mesh-type part of the filter which traps large dust particles is the most important part of the filter when you're in radioactive contamination.

Since the filter mesh material will eventually clog up as the filter becomes full of "junk" from the air, you can extend the life of a filter by avoiding dusty areas or by improvising a "prefilter" in the form of a piece of cloth or other material taped over the filter intake valve. This cloth will filter out large dust particles so that the filter will only receive smaller particles which will take longer to clog up the filter. The cloth can be removed or the dust tapped off it when you're using the mask, if it becomes clogged up.

Filters can be stored for several decades if care is taken not to expose them to air, moisture, or excessive heat. Even if a filter is exposed to one of these damaging conditions, it will still give a lot of protection from the radioactive dust that might otherwise be inhaled though it will no longer give protection from radionuclides.

Seal filters in an air-tight container which will keep the filters from slowly picking up harmless gases from the air and gradually becoming useless when it's time to extract poisonous radionuclide gases. Don't leave a mask, with its filter in place, exposed to the air. The filter will gradually go bad. If you have to keep a filter in your mask (which is practically a necessity in some masks which have hard-to-change filters), then keep the mask sealed in an air-tight bag. (Since moisture is damaging to filters, you must also be careful that moisture doesn't condense inside the bag and get into the filters.)

Industrial respirators are also available and offer a less expensive alternative to more expensive masks. Many have integral filters while others are capable of using filters designed for full-face masks. Respirators are only half-masks that go over the mouth and nose, with filters built into them (usually on the cheeks). The problem with these masks is that they do nothing to protect the eyes. But it's possible to wear air-tight goggles with them to protect your eyes from harmful vapor (as is done with the suit kits mentioned above). Such an arrangement often offers a nice savings in money but does create a system that takes extra time to get on and makes decontamination harder since dust can settle on your face between the goggles and mask.

Probably the best goggles to use with a respirator are the Willson goggles (product number 8-00746) from Industrial Safety and Security. Current retail price for these goggles is $14.50. The goggles are unventilated so that they keep radioactive dust out of your eyes.

Willson, 3M, and North all make excellent half-mask respirators.

North offers a respirator that accepts the filters designed for their full-face mask (covered below). This respirator is available from Direct Safety and costs $19 (without filters which you'll need to purchase for it).

Willson also offers the 1200 series respirator. These accept the filters designed for their 1700 mask (listed below). Cost for this respirator is $15.

3M offers disposable masks which would be good for one-time use in a nuclear emergency escape situation. A number of these masks might also be used for long-term decontamination work. The 3M half-face masks are actually just filters with straps to hold them over the mouth and nose. Because the masks cost from $10 to $13 each, however, it's generally a better buy to just purchase a Redi-Safe kit (listed above); it contains a 3M mask and gives you a whole disposable suit for a few dollars more than the cost of a single mask.

As mentioned before, for heavy radioactive contamination, a full-face mask is much better than a half-mask. One good industrial mask is North' s Full-Face mask which uses a pair of filters (there's one in each cheek of the mask). The mask has a full-face lens to allow good vision when wearing the mask. The North Full-Face mask (number N-03-553) costs $148 and is available from Direct Safety. (In addition to the filters for the mask, it would be a good idea to purchase the nose cup assembly which redirects the air coming into the mask across the eye piece of the mask so that it's less likely to fog up. The nose cup assembly costs $42.)

The best filter set for the North mask when using it for avoiding nuclear contaminants is the 7500-81 dual element filter. Each of these is actually two filters, one over the other. The top filter removes dust and the second absorbs radionuclides (as well as many fumes and organic vapors). These filter sets are also available from Direct Safety for $15 per pair (remembering that you need two for the North mask).

Willson also makes several excellent industrial masks (as well as those for the military). The Willson masks are available in three designs: masks with "in cheek" filters, "chin-style" masks (one large filter in the chin); and masks with the filter connected to it via a long tube with the filter worn on the chest. Unfortunately, filters designed for use with radionuclides are not readily available for the Willson chin-style or chest filter masks; this pretty much rules them out for use in a nuclear emergency.

Willson's cheek filter-style 1700 mask is less expensive than others and quite comfortable to wear. But the filters for the mask are smaller than those for other masks and don't last as long as many other filters. Another problem is that the mask comes only in a regular size so that it won't fit smaller faces. Cost is $104; the mask is available from Industrial Safety and Security.

For dealing with radioactive contaminants, the most useful cheek filter for the AR 1700 mask is the R12 filter. This is designed for use with dust, fumes, mists and radionuclides, making it ideal for nuclear emergencies. These filters cost $20 for a box of 3 pairs. You'll also need to buy two retainers to hold the filters in the masks; retainers for the AR 1700 cost $2 per pair (part number R682).

The Willson 1700 mask will also accept a spectacle mount for those who need glasses to maintain good vision when wearing the mask. The spectacle mount costs $10.94 and fits the Willson mask. A local optician can then mount the prescription lenses into the spectacle mount.

Military masks available on the surplus market offer some of the best buys in the way of masks suitable for use in a nuclear accident. Some masks are much better than others, however.

When purchasing an older or used military surplus mask, check the rubber to be sure it hasn't started to rot. Pull on a rubber strap or the mask itself and observe how it stretches. If you can pull lightly on it and see small cracks in it, then the rubber is rotting and the mask may fail. Such a mask is good only to a collector. (Be sure to test new masks from time to time this same way to be sure they're still good. Storing a mask in an air-tight container will minimize the aging of the rubber.)

Older masks available on the military-surplus market may only have training filters or filters that have been exposed to heat or the air so that they are all but deactivated. These filters still will remove nuclear dust from the air the user breathes and are therefore better than nothing and would be useful in a pinch. However, unless you can get new filters for a mask, it won't be giving you full protection from nuclear gases which might get through a filter that isn't working.

Since pre-WWII filters weren't designed to deal with radionuclides, many old masks aren't suitable for use in nuclear contamination unless they will accept more modern filters. Thus, British CD masks, WWI- and WWII-vintage masks, and the like, will only give protection from dust but wouldn't be ideal for use where nuclear contaminants are involved.

The M9A1 mask is good, but many are getting old and filters are hard to find for it. Because it is used by many European countries, however, adapters are available to allow it to be used with new NATO filters.

One good mask that has been all but phased out by the military is the US M9A1 (with a cheek filter on the left side of the mask). Although ideal for nuclear emergency use, the M9A1 is becoming hard to find filters for. However, some European military units have been using the mask and an adapter is available from many European industrial suppliers which allows the use of the standard 40mm-size neck on filters. With one of these adapters, a wide range of modern European filters can be used with this mask.

Two other US Military masks which are good but have very hard-to-obtain filters are the US Navy Mark V mask and the US Air Force MCU 2-P full-face mask. Possibly these might be modified to use industrial filters but, chances are, you'll be better off with some of the other masks listed in this chapter and leave these two masks to the collectors.

The extra hardware on the front of the M17A1 mask (left) often develops leaks. This makes the M17 mask (right) a much better choice for nuclear accident use.

The M17/M17A1/M17A2 masks are currently being replaced by US Military by the newer M40 mask. Consequently the M17 series of masks are currently readily available on the surplus market and filters are still easy to obtain. Of the various models in the M17 family, the original M17 mask is the best; it doesn't have the drinking and resuscitation tube connectors in the nose pieces which often develop leaks.

The M17 mask has good points and bad points. The mask's filters are inside the cheeks of the mask which makes them less apt to catch on things when the mask is being worn but also makes the filters very awkward to replace. Too, the filters are small so that they won't last as long when exposed to contaminants as the larger filters of most other masks. Military eye-glass inserts are also available for the M17 masks for those who wear prescription glasses and the combat glass frames listed below fit into the mask as well.

Two companies which offer good prices on the M17 masks are Parellex and Sierra Supply (Sierra carries a number of filters, accessories, and spare parts for the mask). Prices will vary with the supplies of masks available, but currently the price of a good mask is in the under-$70 range.

Most M17 masks will come with training filters in them. With the M17 series of filters, the training/tear gas filters have a black ring on them and are coded as the M13: 4240-00-678-8474 and M13A1: 4240-00-934-7854. Filters which will offer protection from nuclear contaminants (as well as from CB agents and many dangerous gases) are the yellow- or green-ringed filters. The green filters are the newest so they would be preferable to the yellow. (As the masks have been phased out of use by the military, they are actually all-purpose.) Yellow-ringed filters are coded:

"M13A1: 4240-00-152-1607" and green-ringed filters are coded as "M13A2: 4240—00-165-5026." To see the color ring, remove the filter cover cap on the outside of the mask. New filters are available from Sierra Supply for $13 for the yellow-ringed filters and $10 for the green-ringed.

Another surplus US Military mask currently available is the M-25 tanker's mask. This mask has a wide, "panoramic" flexible vision plate which gives a wide field of view. The M-25 has a large filter which is connected to the mask via a hose and that is carried in its gas mask bag. While a little awkward, this does give the user a large filter while keeping the weight on his head light. Also, the tube allows the mask to be adapted to commercial filters (with a little work and an improvised clamp). The M-25 is available from a number of surplus sources including SI Equipment, Ltd. (for $40). SI also sells spare filters for $10 each.

Currently the US M40 mask isn't available on the surplus market but it probably will be shortly. It features a filter that is quicker to change and a panoramic lens which gives much better vision to the wearer.

Foreign military surplus masks are also available in the US. One of the best bargains at the time of this writing is the Israeli Civilian/Military mask. Most of these masks were made in West Germany and — though the eye pieces are a little small — are well-designed. The masks come with filters which were designed for chemical weapons but will also absorb radionuclides and the built-in prefilter will remove biological or dust particles. In short, they're ideal for a nuclear accident (as well as defense against chemical and biological weapons). The best part is their price; because there is little market for surplus masks and Israel is replacing the masks with new masks that have larger eye pieces, they've practically given the masks away to dealers in the US. Because of the competition between dealers, the price has fallen from over $50 to less than $10. At the time of this writing, Sierra Surplus is currently selling single masks (which are in new condition) with a new filter for $7 each!

The Israeli masks have a large but light-weight filter that screws directly into the nose of the mask, which makes it easy to quickly replace the filter — something that can't be said about many American masks. The filter also uses the common NATO thread so new Israeli and many European filters fit it; finding spare filters won't be a problem in the near future. (This also makes it possible to use the mask with different filters designed for chemical accidents or the like.)

Most Israeli masks also come with a plastic insert which can be screwed into the mask; this restricts the intake of air when the mask is worn to simulate the restriction created by a filter. This allows a mask to be used in practice without ruining its filter.

The Israeli masks are being replaced by two new masks. One has twin triangular lenses similar to those of the M17 (and can be used with eyeglass inserts). The other model has a panoramic lens similar to that of the US M40. Both masks use the same standard 40mm NATO threading, so obtaining filters for them is simple and, like the Israeli masks they replace, the masks accept the practice training inserts, so the mask can be used for practice without ruining a good filter.

Currently brand new masks of both versions are being sold as "M30" masks here in the US. Both models are available from Brigade Quartermasters with practice insert and a new filter, for $100 each.

(The "old" Israeli filters also fit this mask, so it can be used with the other Israeli surplus masks.)

The Israeli masks are excellent buys in today's military surplus market. Their standard-size filters mean that owners will have a ready supply of commercial filters — as well as surplus filters — to purchase for use in the masks.

Regardless of what mask you choose to purchase, if it doesn't fit properly, it can't give you full protection. Therefore, it's wise to purchase the proper size mask when possible.

Modern masks used by Western European countries, the US, Israel and others come in three standard sizes: small, medium, or large. The size is inscribed on the front of most masks, on the right temple; "S" for small, "M" for medium, or "L" for large.

Unfortunately most mail-order masks come only in medium size. This is not much of a problem for most users and most modern masks are designed so that their edges mold somewhat to a wearer's face to allow some leeway in fit. But a correctly fitting mask is a whole lot better. Therefore, whenever possible, try to order the size of mask that will fit your face (hat size — small, medium, or large — will generally get you into the ball park on sizes).

To get most masks on, you must first put each of your thumbs under the lower inside head straps and pull the mask open and up over your chin and over your face while your thumbs pull the straps back over your head. Do *not* try to place the mask over your head and then down. Unless you're bald, you'll get your hair tangled in the straps and ruin the mask's seal to boot. And you'll probably feel as if you almost ripped off your nose as well.

To test the fit and seal of a mask, first put it on and then put your hand (or hands if the mask has two filters) over the inlet valve(s) on the filter(s). Now try to inhale. If the seal is good, the mask should collapse slightly and remain air-tight for ten seconds while you hold your breath with your hands over the inlet valve(s).

If the mask leaks, either the mask is improperly fitted, you have hair breaking the seal along the edge of the mask, or the mask is malfunctioning. (You can also test the seal with a gas or spray aerosol. Test ampules for testing masks are available from Direct Safety Company; a package of ten ampules costs $6.)

If the mask leaks, first check the mask. The exhaust valve (usually in the mouth/nose piece of the mask) will sometimes be stuck open especially if the mask has been in storage for a while. The valve can generally be unstuck by inhaling/exhaling very hard with the intake valves covered with your hands.

If this valve is working, try to locate the position of the leak and then check the mask to see if there is a tear in it.

If the exhaust valve and mask appear to be in good condition, check the edges of the mask to be sure hair isn't holding the mask away from your face. Sometimes just a small strand will cause the mask to fail (and beards rarely allow a mask to work well).

The fit of the mask can be checked by looking in a mirror (or by having someone inspect you while you wear the mask). The mask should not force your eyes partly closed and none of the straps should touch your ears. If the mask has a nose cup inside it, the cup should not touch your nose. The mask should extend well up onto the forehead and should not cut into your neck under your chin.

If the mask doesn't fit, the problem may be in the adjustment of the straps rather than the size, so try adjusting the straps before giving up on the mask. Try tightening the mask if it's too small or loosening straps if the mask appears too big. (On US masks, the straps can be tightened with a quick jerk back on the end of the strap or loosened by pulling the end forward and, with the other hand, pulling the strap behind the buckle back. On Israeli and many European masks, straps are tightened by pulling the strap back or loosened by pulling forward on the lock release.)

If you need, but can't find, a small mask or need an extra small mask for a child, it's possible to make cuts in the rim of the mask and overlap them by sewing the edge together. Plastic rubber or similar material is then used to seal the mask. This is not nearly as good as having the proper size mask but is better than nothing.

It's also possible to obtain masks for children between 3 and 8 by getting the A-62 (Size 5) mask which is available from many European sources of equipment (see Appendix).

Spare parts can often greatly extend the life of a mask and save a lot of money as well. Since some parts of masks tend to wear out before others, a few spare parts can often make one mask with spare parts last longer than several masks without repair parts. (With the inexpensive prices of the surplus Israeli masks, however, it's just as cheap to purchase several entire masks for spares as it is to purchase parts for them. But with most other masks, this is not the case.)

When possible, if you purchase a number of masks for your family, it is wise to get the same model of masks so that one spare parts kit can be used to service all the masks. Critical parts to have are the exhaust valve, the inlet valves for either side of the upper nose cup, and spare inlet valves (if your mask uses them between the filters and mask).

Valves in the mask should lie flat; if they don't, they'll probably leak. If the small rubber valves in the nose cup leak, the valves can be covered with tape in a pinch so that the eye pieces don't fog up. (It's much better to have spare valves, of course.)

When you're wearing your mask, you may find that the eye pieces fog up inside. This often happens if the nose cup valves are not working properly (or if your mask doesn't have a nose cup). One solution to this problem is to purchase anti-fogging chemicals to put over the lens of the mask. You can purchase commercial anti-fog material either as a stick ($3.95) or in a spray bottle.

These chemicals are usually available in ski supply stores or can be purchased from Brigade Quartermasters for under $4 per container. The chemicals pretty much do away with fogging if used regularly.

Masks like the M17 with eye lens "outserts" or replaceable lenses would be good to have spare outserts for. Since these can become scratched up, visibility and the useful life of the mask will be extended with spare outserts.

When you actually need to don a gas mask, it's wise to hold your breath while putting the mask on and then exhale every bit of air that you can push out of your lungs once the mask is on. This long exhalation of air will push contaminated air out of your mask. Whenever you practice with your mask, do this so that it becomes a habit.

When no one's around to think you're an invader from Mars, try out your mask to see how it obstructs your action and vision. Try running, sitting in your car, and doing a few odds and ends around the house to see just how awkward things become when you're using a gas mask. While masks can save your health and maybe even your life, they slow you down and this should always be remembered when making any plans. Extra time is needed to carry out many simple tasks when you're wearing a gas mask.

To clean a gas mask it's generally sufficient to swab it off with a cloth and soapy water. For a thorough cleaning, the filter should be taken out and the whole mask should get a 4-minute bath in warm, soapy water. Take care to rinse the mask off in clear water for 4 minutes afterwards so that the soap won't irritate your eyes next time the mask is worn. After rinsing, the mask should be dried off carefully because any moisture will damage the filters. (If the mask has been exposed to fallout contamination, take care to dispose of towels used in drying the mask as well as the water it was washed in since they will be contaminated.)

Rubber gas masks normally extrude a whitish material which may form a light film on the surface of the mask. This material protects the rubber; not cleaning this white film off will help the mask last longer.

Some filters like those of the Israeli masks are easily stored. Just put a piece of duct tape over the openings in the filters so air won't flow through them. The filter can then be quickly brought into action by ripping off the tape and screwing the filter into the mask. M17 filters or others which are awkward to put into the mask should be left in their sealed wrappers or placed in the mask and the mask sealed in an air-tight container.

Water can also damage a filter that's in use. Take care to keep out of rain or droplets of water that may result during decontamination procedures. Cold weather can also create moisture problems of a different sort with a gas mask. This occurs when the moisture from your breath forms ice in the outlet valve of a mask; such ice can keep the valve open so that air can leak back into the mask through it.

The way to prevent this is to purchase a cold-weather kit for the mask (which is available for most military masks) or by improvising one in the form of a sock-like piece of cloth which is placed over the outlet valve. All that is needed is to slow down the exit of the body-heated exhaled air so that the heated air around the valve will keep it from icing up.

If a filter should get wet, become damaged, or start to create problems when you're breathing through it, then it is time to replace it. If you can't get to an uncontaminated area to replace the filter, you'll risk contamination if you try to change it in the field. However, in a scrape, it might be better to try to hold your breath and quickly change a filter than to continue with a malfunctioning filter in a radioactive environment.

With some masks (like the Israeli and many European industrial masks) this change can be done quickly; with others, like the M17 series, it would be next to impossible.

If you have a mask whose filter can quickly be changed, then it makes sense to have a spare ready to go at all times; with hard-to-change filters, a spare is optional (though ideally you would have one available).

When disposing of a filter that may have radioactive contamination in it, remember that it may contain dangerous amounts of radioactive material and dispose of it accordingly.

As mentioned before, some models of masks have eyeglass inserts which allow prescription lenses to be placed in the masks. If you wear glasses, it's wise to purchase such a mask. (You might get by with contact lenses inside a mask in a nuclear environment, but many who wear contacts find the air coming through the mask is somewhat irritating to their eyes.)

For the M17 and M17A1/A2 masks, the US Army created wire frames for prescription lenses. A "plastic optical insert" only for the M17A1 mask can also sometimes be found. (Sherwood International currently carries these.) Once you purchase the frames, an optical store can fit them with prescription lenses.

Optical inserts designed for scuba masks work well in many gas masks (shown here is the M17 mask). These frames are available at most optical stores or can be ordered from Brigade Quartermasters.

With some masks it's also possible to use the new black nylon sports glasses that are designed to be used with scuba masks. These frames are held in place with a rubbery plastic strap that is thin enough to allow a gas mask to seal around them. These are available through most optical stores or may be purchased directly from Brigade Quartermasters for $25.

These frames are a tight fit in most masks and may be impossibly tight if you wear a small mask. A little slack can be obtained by ordering the small size of the frames (regardless of the size of the glasses you normally wear); the outside corners of the frames can also be trimmed off slightly with a knife or file without making the frame too weak. These frames will work in the Israeli masks, the M17 series, and in most industrial masks.

If you have an old pair of glasses, you can make an optical insert by removing the ear pieces, trimming down the outside edges of the frames, and fastening them into the mask. It would also be possible to remove the lenses from the frames (soak them in boiling water to get the plastic to give a little before attempting this) and epoxying the lens inside the lens of the mask. While neither of these methods are very satisfactory, they are better than nothing for those with major vision problems.

The mask should go into your bug-out bag, or have some sort of carrying bag so that you can have the mask with you in an emergency. Having a mask off somewhere you can't get to it in a hurry will be of little help to you. Currently there are large numbers of M17 and M9A1 bags which will work well with almost any gas mask. Brigade Quartermasters and other companies also have a number of carrying bags suitable for such use. If you store the mask with the filter in it, keep the mask in an air-tight ziplock bag.

If you have a hooded Tyvek suit or similar outfit to go with your mask, then a hood designed for your mask won't be a necessity. A hood does give some extra protection and might be a useful accessory for a mask, however, since it would make it a little less likely that dust will become lodged around your face. Furthermore, a hood will stop some minor leaks around the edge of a mask. And if you're using some type of suit without a hood or are "improvising" a suit from clothing, then a hood would be a necessity.

Many gas masks have hoods designed for them. Shown here is the M17 mask with its hood in place. (Note duct tape used to seal zippered front of coveralls and the areas between the gloves and shirt sleeves.)

Many gas masks have hoods designed for them which can be purchased from military surplus stores. Hoods designed for the M17 series of masks are currently the most readily available on the surplus market (Sierra Supply sells them for $5 each). The

M9A1 and M25 tankers' masks also have hoods which occasionally can be found in the surplus market.

The rubberized M17 hood can also be adapted to a number of masks since it is readily cut with scissors and reglued with contact cement or the like. If the lens openings of the M17 hood don't fit around the new mask, it's possible to glue clear, heavy plastic in the eye holes of the hood with contact cement. Seal any possible leaks in the hood with plastic rubber or "Goop" (a rubbery product available from Industrial Safety and Security Company for $3.33). This material can also be used to repair tears in hood materials.

(To get the M17 hood on an M17 mask, the "outserts" of the mask's eyepieces are taken off and the hood pulled over the mask and the black line over each eye hole is lined up over each eyepiece. Next, hood holes are then stretched around the eye lens, the filter caps, and the nose piece (be careful not to rip the fabric of the hood). The drawstring of the hood is then tightened around the nose piece, tied in a double knot, and tucked inside the hood. With the replacement of the lens outserts the job is finished.)

When putting a mask and hood on, use the same procedure outlined above for putting on a mask but turn the hood wrong-side-out over the mask before you start. Once you get the mask on, pull the hood back over your head and tighten the drawstring around the outside of the hood. If your hood has cords to hold down its edges, fasten them under your armpits to hold the hood down.

A complete suit will protect you from alpha and beta radiation and simplify the removal of contaminants when you clean up and enter a safe area. The less time you spend in a contaminated area, the better. Get out of, or avoid, contaminated areas when possible. Use the suit and mask to help you leave dangerous areas or to carry out essential decontamination work.

DECONTAMINATION

If you are trapped in the open when the fallout from a nuclear accident starts to arrive, your first task is to get out of the contaminated area as soon as possible.

Those very close to where the disaster takes place may actually be able to see particles of material which may be radioactive as they fall back to earth. These particles can take almost any form depending on the type of accident involved and could be sand-like, look like soot, or be small chunks of material.

If you are in such an area, you should try to keep the fallout off your skin by covering your skin and by brushing radioactive dust off your clothing and skin. If possible you should wear a Tyvek suit, mask, goggles, etc. If you don't have such equipment, at least cover your head and use a wet handkerchief over your mouth and nose to keep from inhaling the dust. The less time spent in the fallout, the greater your chances for maintaining good health.

After traveling through a contaminated area, even if the contamination is invisible to the naked eye, you should take all the precautions you can by brushing off the clothing you're wearing so that fine dust and particles won't be taken in with you to a relatively contamination free area.

Since it isn't possible to instantly decontaminate yourself, you'll have to clean up in stages. This will also create areas which are somewhat contaminated where you engage in the decontamination. Decontamination areas should be large so that this clean up can be spread out so that you move from greater to lesser areas of contamination as you decontaminate yourself and your clothing.

The first step in decontaminating yourself is to brush yourself off very thoroughly so that any radioactive dust which may have settled on you is removed. This done, you should move to a clean area and remove any tape which has been used to seal up your clothing. Next, move to another cleaner area and remove your clothing, hat, boots, Tyvek suit or whatever and toss them into a "contaminated materials" area or in large plastic bags. Remember that the less you move contaminated clothing about, the less dust will get into the air.

Once your outer clothing is off, you should remove your dust mask, goggles, and gloves. Hold your breath as you do this so that radioactive dust isn't inhaled. Discard the mask and gloves in another area or put them into a plastic bag and seal it for later decontamination if you'll need to use the equipment again.

To decontaminate yourself fully, you should take a shower if at all possible. Don't use a sit-down bath for decontamination as this won't do a thorough job. If a shower is impossible, at least wash your hands and face with copious amounts of water. When washing, pay special attention to cleaning under your finger nails and toe nails and all folds and body creases. When possible, wash your hair and body with soap, shampoo, or detergent. (If you have long hair, trimming it short will simplify decontamination procedures and minimize chances of retaining radioactive dust in your hair.) Decontamination of your skin can also be carried out by using chelating agents (chemicals — developed for the nuclear industry — which bond with radioactive particles so that they are easier to remove). These can also be used for decontaminating open wounds.

Chelating solution is sold under the trade name of "Radiac Wash." It is available from Dosimeter Corporation and Guillory and Associates (the latter offers it in 1 liter bottles for $15 or in pre-moistened towelettes with a box of 100 being sold for $20.) Both the liquid and the towelettes are effective in decontamination procedures. Chelating agents are especially effective when used in conjunction with the shower clean-up routine above.

After showering, if possible, a low-level radiation meter should be used to search for any contamination you may have missed when washing. If any radioactive particles are indicated, carefully wash the area again until the contamination is removed. Discard rags, towels, or towelettes which have been used in cleaning up since they will be contaminated.

Once you've evacuated an area that is heavily contaminated, you'll probably not be able to re-enter it during your lifetime because of the long half-lives of many radioactive elements involved in the nuclear industry. Likewise it's impossible to "tough it out" in a shelter the way you might during a nuclear war and the radioactive fallout; because the half-life of most nuclear accident material will be so much longer than that of fallout from a nuclear bomb, you couldn't outlive the radiation. If there's a reactor accident or some similar major accident, plan on evacuating to a safe area.

If traffic allows you to evacuate an area in a car, while you're in the contaminated area, keep car windows closed and wear a protective suit. If possible, keep your protective suit on until you can decontaminate yourself in a non-contaminated area. If the temperature becomes warm, use an air conditioner rather than opening car windows while you're traveling in a contaminated area; use the "recycle" mode on the car's air conditioner rather than the "fresh air" selector to minimize radioactive particle intake.

Once out of the contaminated area, the car should be washed and decontaminated, then checked for radioactive particles before it is used again.

If you're forced to stay in a lightly contaminated area, you can carry out some decontamination procedures to minimize your risks. Stay inside as much as possible and keep the house sealed up to minimize the amount of radioactive dust which gets into the house. Vacuum sweepers should be used (if you have electrical power in the area), and dusting should be carried out frequently to remove contaminants from various surfaces inside the house.

Rain and the wind will gradually disperse many dangerous radioactive materials into low-lying areas and — eventually — wash much of the contamination into rivers and lakes where it will settle down to the bottom. Once in rivers and lakes, the contamination will be relatively harmless *provided* fish from the river aren't eaten or the water used for drinking without proper filtration.

If you are forced to stay in an area which has light contamination, it may be hard to get food and other supplies for a short time since many truckers and other workers will be fearful of entering the area. Because of this, it is wise to always have extra food in your home. This can tide you through a number of emergencies other than a nuclear accident. (If enough people would do this, relief work during emergencies would be greatly simplified. Too, large numbers of people purchasing extra supplies for emergency use would help out the depressed farm economy. While many people fail to prepare to supply themselves during emergencies because of fear of "hoarding," purchasing supplies when they are abundant is *not* hoarding. It's "stockpiling" and it makes good sense for those living in an interdependent age like ours.)

Low levels (or even very high levels) of radiation won't harm food. All that needs to be done is to remove the radioactive contamination from the food so that you don't ingest the particles. This is simple with food in sealed containers; just carefully wipe off any dust (and possible contamination) on the container before opening it.

Processed foods do have a finite shelf life. Once this time is exceeded, the nutritional value of the food gradually drops off. Food will remain eatable for some time but it may not meet many of your body's nutritional needs. Because of this, it is wise to purchase extra food for your "emergency stockpile" and then rotate it with the food you use so that it is eaten before its shelf life is exceeded.

Processed foods have the shortest shelf lives. Most canned foods in cans or jars have shelf lives of 6 months. Canned meats and non-citrus fruits will last longer than other canned foods and will have some food value for up to a year.

Evaporated milk has a shelf life of 6 months. Cereals, bouillon, nuts, instant cream, and hydrogenated (or antioxidant treated) fats/vegetable oil also have nutritional shelf lives of about a year.

A few processed foods last a longer time; coffee, tea, cocoa, candy (that's nearly 100% sugar), spices, sugar, salt, pepper, etc., will last many years. The catch is that most of these have little nutritional value. If you want to extend the shelf life of any of these, it is possible to add a month to them by storing them at low temperatures.

Freeze dried and dehydrated foods have shelf lives of 4 to 6 years; they don't store perpetually as many advertisers claim. Dried beans, soybeans, wheat, and other grains have shelf lives of four or five decades; corn, if kept cool and dry, will store for two decades. The catch with these unprocessed foods is that you must have some way of milling them and go to a lot of work in preparing them. However, if you wish to simply purchase the supplies and set them aside as an "insurance policy" against a shaky future, then this is often the simplest way to go.

Hunting or trapping might work for putting food on your table if you live in a remote area where hunting is practical. But it is a good way for the person doing the hunting to be exposed to extra radiation and the meat may be a source of extra contamination as well. As we'll see, however, it is possible to process meat from contaminated animals. If the choice is starving or risking the possibility of getting a radiation-induced illness 30 years down the road, the choice shouldn't be hard to make. Likewise, it would be possible to forage for eatable plants in some areas, provided there weren't large numbers of people competing with you for the food.

The bottom line is this: don't eat food that may be contaminated unless you have to. If possible, eat food that has been in sealed containers or has been brought into your area by relief workers.

But if you have to eat food gleaned from contaminated areas, it is possible to process it to minimize your risks (a good low-level meter will also help you to check your decontamination work when processing such food).

Plants can be processed most easily when they have been in an area of radioactive contamination only a short time. In such a case, radioactive particles can easily be removed from them. As time goes on, plants will incorporate radioactive substances into their cells so that it will be hard or impossible to decontaminate them. Thus, the longer a plant has been in a contaminated environment, the less safe it will be to eat even with careful processing.

To process plants, carefully peel and clean the plants... that's all you need to do. Since most of the initial contamination will be removed with the outer layers of plant, you will be able to eat without fear of ingesting much radioactive material. Plants with smooth skins (like tomatoes or green peppers) can be just washed off. With plants like potatoes whose edible parts come from the ground, you will be less apt to have the plants become contaminated if you first remove the top layer of soil around the base of the plant before digging the edible roots or tubers up (they still need to be carefully washed and peeled, of course).

Domestic or wild animals can also be processed. Again, the sooner the animal is harvested after the initial contamination, the safer it will be to eat.

By observing the animal's behavior when it is still alive, you can also gain some clues as to whether or not it should be eaten. If it looks healthy, it would probably be safe. A sick animal, on the other hand, may indicate that the beast has been exposed to high levels of contamination. Even if the animal is processed, it may contain dangerous toxins created by bacteria or viruses. (Since radiation lowers an animal's resistance to disease, an animal's sickness may be indicative of high radiation exposure.) Cooking will kill bacteria or viruses but will *not* destroy toxins. Eating meat from a sick animal will poison you.

Animals which may have ingested radioactive contamination but still are healthy could be eaten in an emergency. Since animals' bodies tend to store dangerous radioactive material in certain locations in their bodies, avoiding eating these areas will greatly reduce the risk of ingesting contaminants . The parts and areas to avoid eating include the thyroid glands, the kidneys, the liver, and meat next to the bones as well as the marrow in the bones themselves. If you avoid eating these parts of the animals and eat only muscle meat, you'll greatly reduce your radiation intake.

You should also be sure to thoroughly cook the meat so that *all* bacteria are killed in the meat. Again, since radiation lowers resistance to disease, the animal may have high concentrations of bacteria in it. And radiation will have lowered your resistance to disease as well, so cooking the meat will lower the amount of "fighting" your body has to do to stay healthy.

Remember that the waste parts of both plants and animals which you shouldn't eat are probably contaminated. These parts should be disposed of in a manner that will keep them from contaminating your water or crops.

Animals which are sick of radiation exposure may recover so don't kill off any animals which are ill. At the same time, because the resistance to disease of most animals will be lowered, sick domestic animals in a herd or flock should be separated from the others in the group so that the disease can't spread.

Farm crops which were in a contaminated area could also be made safe to eat. Wheat, corn, and most other grains would be fairly well decontaminated by the processing necessary to free the kernels from the husks. Therefore, crops exposed to contamination would be relatively safe to eat (though food from non-contaminated fields would, of course, be even safer).

If the contamination were allowed to sit on fields while the plants were growing, however, the situation might change. In such a case, it is probable that radioactive contamination would be incorporated into the plants' structure making crops dangerous to eat from a long-term health stand point. (Of course, if the choice were between eating the crops or starving, obviously it would be better to risk eating such crops.)

Getting drinking water as well as water for personal decontamination is also possible in areas which have been lightly contaminated. Fortunately, contaminated water is seldom so contaminated it can't be cleaned up and used.

Since local water works (as well as power and other utilities) may go "off line" during a nuclear emergency, you'd be wise to have secondary sources for water. You should locate springs, wells, streams, etc., in your area *now*, before an emergency occurs.

Another good idea would be to store water which you can fall back on during an emergency. Water stored in sealed containers can simply have its containers dusted off, some water purification tablets dropped in (to kill off any algae or micro-organisms which may have taken up residence in it), and it will be ready to use.

How much water will you need to store?

A human being should have at least a gallon of water per day during hot weather or 2 quarts per day if the weather is cool; this is if the person doesn't do any heavy work, and no water is needed for cooking or cleaning. The 2-quart-per-day amount is a *minimum*; a gallon per day is a safer amount to plan on, however. Figure out how many you wish to have water stored for and for how long you wish to be able to supply yourself with water and then store that many gallons.

When storing water, remember that it's heavy; don't store it where its weight may cause structural damage to your home or where you'll have to carry back-breaking containers about before they can be used. Storage containers can cost a lot of money or be very cheap.

If you purchase commercial storage containers, your cost can be 3 to 10 dollars per gallon. That's expensive water!

Fortunately, if you're like most people in the US and elsewhere, a lot of potential water storage containers are available to you in the form of tough plastic soft drink and milk containers. A second good source is available in the form of used food containers available from Dairy Queen stores, restaurants, etc., for free or a small price. With a little clean up, these are ideal for storing water. All you need to do is to be sure the container was designed for food; don't use plastic which is not designed for food storage since it may contain chemicals which will leach into water which is stored in them. (My favorite containers are the two-liter pop bottles. These are tough and easy to handle. Plastic milk containers work, too, but seem prone to leakage if they are abused in the least bit.)

Don't use glass containers as these can be quite dangerous or may break under the rough-and-ready conditions of a disaster. Things are bad enough during a nuclear emergency without having broken glass scattered over your living area and your water supply suddenly down the tubes, as it were.

glass scattered over your living area and your water supply suddenly down the tubes, as it were.

Containers larger than 2 gallons are hard to handle and the upper practical limit for easy-to-use water containers would seem to be 5-gallons. (When working with any container that's this large, it's a good idea to place it where you want it and *then* fill it; otherwise you risk wrenched muscles or an out-of-whack back.) Remember, too, that any filled container, even a 2-liter pop bottle, is dangerous if it falls from any height. Secure water storage bottles so someone can't be seriously injured by them.

In addition to avoiding plastic containers which are not designed for storing food or water, find out what was in used containers before storing drinking water in them. If they ever contained petroleum products or other poisonous chemicals, avoid using them for storing drinking water. Lethal amounts of hazardous chemicals can be absorbed by the plastic in containers so that there is no way to clean them out.

In theory, metal containers which contained hazardous products can be cleaned out but it is hard work and iffy at best; use only those which have been used for shipping non-poisonous chemicals as well. Metal containers can rust out and the water in them can take on some odd tastes. In general, plastic is much better than metal or glass containers for water storage.

Water taken from areas exposed to radioactive contamination can have the contamination in it reduced, if not completely removed, a number of ways.

Distillation is one way, but is not as good as one might think. Commercial home distillation units are also expensive and require a source of energy to power them. Because some radioactive contaminants are gases or have boiling points lower than water, many will actually follow the water vapor out during the distillation process. Unless complicated fractional distillation methods are used, all but salts and solid particles in suspension will boil out with the water and condense with the "clean" drinking water! Because of this, distilled water can be far from safe to drink.

Water filtration using activated charcoal is probably the best solution to this problem. Commercial filters are still expensive, but they can work well *provided* the flow of water through them is *very* slow and the carbon filter is new. Unfortunately many commercial units are set up in the home water line so that the charcoal doesn't have time to absorb many contaminants from the water and — because of the volume of water going through it — quickly fills up with pollutants. Filters using activated carbon can also become a source of bacteria contamination of water so the water from them should be chemically treated. If you get a commercial unit, use it in the "trickle" manner outlined below and then use the chemical treatment below to sterilize the water.

Simple carbon filters can also be constructed and maintained on a do-it-yourself basis. While it's a bit more work to make such a filter, it can be kept going for some time and is flexible enough in design to be made from odds and ends in a disaster.

The heart of a carbon filter is activated carbon (also called activated charcoal). This carbon material is created from coal and wood which has been processed at high temperatures with steam in the absence of oxygen. Activated carbon is very porous and, when broken into bits, has a huge surface area which is capable of "latching onto" a number of chemical compounds so that they become bonded to its surface. Activated carbon doesn't combine with oxygen or many compounds which are formed partly of oxygen (like water); this makes it ideal for a water filter.

On the down side, since carbon works by chemical bonding, it also has a finite life. After a certain amount of water has gone through the filter and contaminants — as well as some harmless chemicals — have been removed, the carbon will cease to work. When this happens, continued use of such a filter doesn't do anything to remove radioactive contaminants from the water. Therefore, the user must be sure to replace the activated carbon in a filter periodically.

Another problem with activated carbon is that bacteria can grow on the surface of the carbon. Because many bacteria thrive on the chemicals bonded to the carbon surface, bacteria growth can be quite rapid. This also dictates that the carbon must be changed regularly, even if little water has gone through the filter. To be on the safe side, it is also wise to sterilize the water after it has moved through the filter.

There are several ways to sterilize the water. One is to boil the water — at a full boil — for 20 minutes. Chemicals can also be used to sterilize the water. Water purification tablets are usually available in most camping supply stores or from Brigade Quartermasters; the best tablets currently available are the Potable Aqua tablets which cost around $4 per bottle of 50. Follow the directions for use on the bottle that comes with the tablets since dosage per quart varies from brand to brand.

Two household chemicals can also be used to sterilize water. One is 2% tincture of iodine and the other sodium hypochlorite — the only active ingredient found in most household bleach. With 2% tincture of iodine, use 5 drops per quart of water; with 5.25% sodium hypochlorite solution — bleach — use 2 or 3 drops. (This also can work to sterilize stored water. Be sure to double the amounts if the water is cloudy or smells bad.)

When using chemicals to sterilize water, the killing of the bacteria is not instantaneous. This means that the water should stand for a half hour before being used.

A final problem with the charcoal filter is that it doesn't stop large particles of contamination. Since large particles of radioactive dust could be found in contaminated water, a charcoal filter being used in a nuclear accident must be augmented with porous material. This material will then remove the large particles that aren't absorbed by the activated carbon. (This filter material will also remove any particles of carbon, which some authorities think may increase the likelihood of cancer in the gastrointestinal tract.)

This mechanical filter section can be created from filter paper, cotton balls, coffee filter paper, cloth, or other material capable of removing small particles that would otherwise find their way through the activated carbon. (Washing the activated charcoal before using it will also remove small carbon dust from it and make the mechanical filter material last longer.)

A good source of activated carbon is a local pet store or five-and-dime store. They sell activated charcoal for use in aquariums. Some garden supply stores may also carry it (it's useful in removing dangerous chemicals from the soil).

If you are unable to purchase charcoal, it's also theoretically possible to make activated carbon by placing wood in a sealed container with only a very tiny hole in it and heating it until the wood is charred into activated carbon. If possible a tube leading into the container would be used to force water onto the wood to create steam inside the container, though this is not essential; the main consideration is to not allow air to get to the charcoal when it's being heated, while having a large enough hole in the heating container to keep it from exploding when the hot gases in it expand. All in all, it's a whole lot easier to purchase the carbon before a nuclear accident rather than try to make it under emergency conditions.

Since activated carbon will also pull gases out of the air, it's wise to seal the activated charcoal in an air-tight container or plastic bag until it's needed.

Constructing an activated carbon filter is simple. If you have a large funnel or small plastic jar which you can poke holes into the bottom of, all you have to do is line the inside of the container with the mechanical filter material which will stop the large particles coming through the filter (a number of materials will work for this including coffee filter paper, cloth, cotton balls, etc.). Next, fill the inside of the filter with the activated carbon. And your filter is finished.

To use it, drip water slowly into the top center of the activated charcoal. As the water slowly moves through the carbon, it becomes purified and drips out the bottom in a purified state. Boil or chemically sterilize the filtered water and it's safe to drink. (If you won't be drinking the water for a while, store it in a cool place, such as a refrigerator.)

Just remember to keep the water running very slowly through the filter and to change the carbon every 3 weeks or after 20 gallons of water have been filtered by it. When changing the charcoal in the filter, it's also a good idea to disinfect the inside of the tubing. The best chemical to use for sterilization is straight bleach (sodium hypochlorite); then let the container stand and dry afterward. Dispose of used container stand and dry afterward. Dispose of used activated charcoal carefully since it may be dangerously contaminated.

A larger filter could be created with a section of plastic pipe or metal tubing. While the most durable filter could be created with copper tubing and galvanized pipe, the easiest to build is constructed with plastic pipe. Three or four-inch diameter plastic pipe is easy to work with and is probably the best material for such a purpose.

Cotton balls (or other filter material) can be used in the bottom of the filter (about 2 inches deep) to keep the carbon and large contaminate particles from getting through the filter. Twenty-eight inches of activated carbon is best for this filter, though a shorter column can be used with larger diameter pipe.

Daily flow rate should be around 6 gallons per day (depending, again, on the amount of carbon in your filter... smaller filters should have a smaller maximum flow than 6 gallons and really big ones could have a greater flow). The activated charcoal should be replaced every three weeks to keep growth of bacteria in it at safe levels, or replaced after 120 gallons of water have been filtered.

Because of the possibility of radioactive contamination on the ground gradually getting into an area's water supplies, it would be wise to filter all water from contaminated areas before drinking it.

Many modern homes are quite "tight" as far as air flow is concerned. This lowers heating and cooling costs of houses and also has the added benefit of lowering the amount of dust that filters into them when the structure's windows are closed. In a nuclear accident which created low levels of contamination, this could be quite important since it would lower the amount of contamination that could make its way into a house.

Since electrical power would probably be available shortly after a nuclear accident in areas of low contamination, it would be possible to "clean up" a house using commercial vacuum sweepers and common household cleaning equipment. Low-level meters should be used to check such work and care should be taken in disposing of dust bags and other potentially contaminated cleaning equipment. (It would also be theoretically possible to seal up a house and create a "filter" from a vacuum sweeper or build one from easily obtainable furnace filters. Provided electrical power were available or could be generated by the user, a home could be sealed up against nuclear contaminants. For a description of such a set-up, see my *Surviving Major Chemical Accidents And C/B Warfare* available from Loompanics Unlimited.)

If large areas of land receive low-level contamination, it is probable that most governments will be unable to actually finance a resettlement program to allow victims to abandon their contaminated homes and land to move to new houses. If any resettlement program is undertaken, chances are also good that the "new and comparable" housing will be greatly inferior to original homes of many of the victims.

Likewise, many insurance companies have limited their coverage during nuclear accidents and many countries have granted their nuclear industries limited liability during major nuclear accidents. In short, the victims of nuclear accidents probably won't receive a lot of financial help following a major nuclear accident. Because of these factors, many victims in areas of low-level contamination may want — or be forced — to stay in place and try to carry out decontamination procedures in their homes and businesses.

Victims can do much to improve things in their area through their own efforts, even without much outside help *if* they have their own monitoring equipment (or help from CD officials with such equipment) and if they know how to carry out such work.

Hopefully, the government of the country in which the accident occurs will also be able to engage in massive clean-up efforts to help the victims. If such is the case, chelating agents may be used which will "eat up" the contamination so that areas can be made nearly free of radioactive contaminants and so that the contamination can't get into water supplies. But even if chelating agents aren't available, it's still possible to remove radioactive particles from the ground so that yards or even farms are safe to live on and even raise food on.

Decontamination of the surface of outdoor areas would need to begin as soon as possible so that contamination doesn't have a chance to sink into the soil. If work is done soon enough, it is possible to remove most contamination by simply stripping off the top layers of plants and soil. Removing the top 2" of soil in any area with undisturbed soil, would remove up to 99 percent of any radioactive contamination which fell onto the area. (Tilled or rough areas would have up to a 50 percent reduction of radioactive dust with a 2-inch removal of soil.) In many areas, this work might not need to be done. On play grounds or garden areas, it would be best to carry out, however.

(Contaminated top soil removed during decontamination procedures would pose a great disposal problem. It would be dangerous and could contaminate ground water if water were allowed to seep through the radioactive soil.)

Farmland or other areas could also be reclaimed by plowing contamination well below the surface. This would allow plants to obtain nutrients from the new surface of the earth while contaminants were under the surface. An added benefit would be that the surface earth would help protect the plants from the radiation by density shielding. Plowing the contaminated top soil so that the surface ended up 18 inches below the surface would reduce the radiation levels on the surface by up to 50 percent.

The intake of radioactive contaminants by plants could also be reduced by increasing the amounts of minerals which are mimicked by nuclear isotopes. For example, calcium could be added to the soil to offset the intake of strontium-90 by the plants.

Again, while it would be better to obtain food grown in non-contaminated areas, it would still be possible to raise food on contaminated ground in an emergency.

If you were forced to live in an area of low-level contamination, you could do much to improve the environment you found yourself in and could even return to life-as-usual if outside help were available to aid in decontamination processes.

THE FUTURE

The short history of the nuclear industry has shown that nuclear energy can be far from "clean." While it is true that nuclear energy can be less dangerous — at least in the short term — than fossil fuels when it comes to the health risks created, it remains to be seen just how safe nuclear energy is over the long haul of many centuries during which wastes created by it remain dangerous.

And short-term dangers can also be great — as Three Mile Island and Chernobyl have shown us.

In the meantime, each of us can do a lot to improve our chances of surviving a nuclear accident while minimizing our chances of both short- and long-term risks.

The future does not look as bleak as one might think. In the near future, it's possible that new, more efficient appliances, as well as conservation practices, may reduce our need for nuclear-generated energy. In the not too distant future, it is probable that the Tokamak nuclear fusion generator may be perfected so that nuclear energy becomes truly clean, inexpensive, and safe. The future may also find that even "old fashioned" fission reactors can be safe; new designs such as the Swedish PIUS (Process Inherent Ultimately Safe) and the US MHTGR, which is melt-down proof, certainly make it appear that safe reactors may eventually replace the current generation of reactors.

It is also probable that today's reactors will be viewed by those in the near future with the same amazement and disbelief as we have when reading about early factories which maimed workers and turned the daytime skies dark with soot.

Nuclear energy will someday be safe. In the meantime, there is much that can be done to improve today's safety standards and to offer greater protection to those who must live in the shadow of a possible nuclear accident and its aftermath.

APPENDIX OF SOURCES OF USEFUL EQUIPMENT AND PUBLICATIONS

There are a lot of publications and equipment available for those wishing to prepare to survive a nuclear accident. Much of the equipment can double to give protection from a nuclear war or even from many chemical and biological weapons or accidents. Therefore, with just a little extra expense, it is possible to obtain protection from a wide range of possible disasters.

With this in mind, some of the publications and equipment listed below may not be related only to nuclear accident survival; rather, it could be used to expand your survival chances from a variety of accidents or war-time conditions.

SOURCES OF USEFUL EQUIPMENT AND PUBLICATIONS

Aegis International (Bi-Monthly Magazine)
Weinbergstrasse 102
PO Box
CH-8035 Zurich, Switzerland
(European CD work and new CD equipment reviews)

Alfred Kaercher GmbH & Co.
Leutenbacher Strasse 30-40
PO Box 160
D-7057 Winnenden, Switzerland
(High-pressure cleaning and decontamination systems)

American Civil Defense Association
Box 1057
Starke, FL 32091
(Publishes the excellent *Journal of Civil Defense*)

American Survival Guide (Monthly Magazine)
Box 2508
Santa Ana, CA 92707
(Survival articles and how-to information)

Andair, Ltd.
CH-8450 Andelfingen, Switzerland
(NBC filters, ventilating units, etc.)

Armscorp of America
9162 Brookville Rd.
Silver Spring, MD 20910
(Military surplus, Israeli gas masks)

Brigade Quartermasters, Ltd.
1025 Cobb International Blvd.
Kennesaw, GA 30144-4349
(Freeze-dried military-style food packets, M30 Israeli masks, "bug-out" bags, and military surplus type equipment)

Brunswick Defense
One Brunswick Plaza
Skokie, IL 60077
(Protective gloves, shelters, etc.)

Choate Machine and Tool
PO Box 218
Bald Knob, AR 72010
(Accessories for Mini-14, AR-15, Mossberg 500, Remington 870, etc., firearms)

Defense And Technology (Magazine)
1965 Broadway
New York, NY 10023-5965
(Monthly magazine which covers nuclear, biological, military and industrial equipment and looks at various aspects of modern warfare and accidents.)

Direct Safety Company
PO Box 8018
Phoenix, AZ 85066
(Retail sales of North, 3M, and other masks and filters as well as other accessories)

Dosimeter Corporation
Box 42377
Cincinnati, OH 45242
(Radiation meters, dosimeters, dosimeter chargers)

Emergency Planning Journal
Emergency Planning Canada
Ottawa, Ontario, Canada KIA 0W6
(Quarterly journal for Canadian civil defense)

Flodins Filter AB
S-45300 Lysekil, Switzerland
(Gas mask filters, filters for shelters)

Guillory & Associates
PO Box 591184
Houston, TX 77259-1184
(Aegis I fallout suit kit (with goggles, mask,etc.), Plessey PDRM 82 radiation meter, "Radiac Wash," and survival publications)

Industrial Safety and Security Company
1390 Neubrecht Rd.
Lima, OH 45801
(Retail sales of 3M and Willson masks and filters as well as other accessories.)

Kirchner GmbH
Tirolerstrasse 85
D-7000 Stuttgart 61, Switzerland
(Protective clothing, acid/radiation protective clothing)

Loompanics Unlimited
PO Box 1197
Port Townsend, WA 98368
(Reference, medical, survival, etc., books)

Louis Schleiffer AG
CH-7814 Fieldbach/Zurich, Switzerland
(NBC protective clothing)

Luwa Ag
Kanalstrasse 5
CH-8152 Glattbrugg, Switzerland
(Filter and protective systems)

Michaels of Oregon
Box 13010
Portland, OR 97213
("Uncle Mike's" inexpensive nylon holsters, belts, magazine pouches, buttpacks, etc.)

Mine Safety Appliances Company
600 Penn. Centre Blvd.
Pittsburgh, PA 15235
(Safety equipment, masks, breathing equipment)

Paladin Press
PO Box 1307
Boulder, CO 80306
(Military, medical, survival books)

Parellex
1090 Fargo
Elk Grove Village, IL 60006
(M17 gas masks and military-style gear)

SGW (Olympic Arms, Inc.)
624 Old Pacific Hwy SE
Olympia, WA 98503
(SGW version of the AR-15 plus spare parts for it or the Colt version of the rifle)

SI Equipment, Ltd.
2322 Artesia Blvd.
Redondo Beach, CA 90278
(Gas masks, freeze-dried/dehydrated foods, etc.)

Sherwood International
18714 Parthenia St.
Northridge, CA 91324
(Military surplus equipment)

Sierra Supply
PO Box 1390
Durango, CO 81301
(M17 masks, Israeli gas masks, M17 filters, accessories for masks, and military surplus equipment)

Tesimax-Altinger GmbH
D-7530 Pforzheim
Holderlinstrasse 39, Switzerland
(Protective clothing and decontamination equipment)

Trelleborg AB
Division of Skyddsprodukter
Box 24 S-271 00 Ystad Sweden
(Protective suits and breathing equipment)

Victoreen Instrument Division
10101 Woodland Ave.
Cleveland, OH 44104
(Radiation meters, dosimeters, dosimeter chargers)

BOOKS FOR FURTHER READING

These books will give you a lot more information on many of the subjects touched on in this book. Most of the publications listed below are available directly from the publisher.

Berkow, Robert, Editor. *The Merck Manual, 13th Edition*. Rahway, NJ, Merck & Co., 1977.

Clayton, Dr. Bruce. *Fallout Survival: A Guide To Radiological Defense*. Boulder, CO, Paladin Press, 1984.

Clayton, Dr. Bruce. *Life After Doomsday*. Boulder, CO, Paladin Press, 1980.

Clayton, Dr. Bruce. *Thinking About Survival*. Boulder, CO, Paladin Press,1984.

Clayton, Mary Ellen with Dr. Bruce Clayton. *Urban Alert!* Boulder, CO, Paladin Press, 1982.

Glasstone, Samuel and Philip J. Dolan, eds. *The Effects Of Nuclear Weapons*. Washington, DC, US Departments of Defense and Energy, 1977.

Klinghoffer, Max, M.D. *Triage: Emergency Care Handbook*. Lancaster, PA, Technomic Publishing Company, 1985.

Lesce, Tony. *The Shotgun In Combat*. Cornville, AZ, Desert Publications,1979.

Long, Duncan. *The AR-15/M16: A Practical Guide*. Boulder, CO, Paladin Press, 1984.

Long, Duncan. *Assault Pistols Rifles And Submachine Guns*. Boulder, CO, Paladin Press, 1986.

Long, Duncan. *Combat Ammunition: Everything You Need To Know*. Boulder, CO, Paladin Press, 1986.

Long, Duncan. *Automatics: Fast Power, Tactical Superiority*. Boulder, CO, Paladin Press, 1986.

Long, Duncan. *Combat Shotguns*. Boulder, CO, Paladin Press, 1987.

Long, Duncan. *The Mini-14*. Boulder, CO, Paladin Press, 1987.

Long, Duncan. *Nuclear War Survival*, fifth edition. Houston, TX, Guillory & Associates, 1986.

Long, Duncan. *Survival Bartering*. Port Townsend, WA, Loompanics Unlimited, 1986.

Long, Duncan. *Surviving Major Chemical Accidents And C/B Warfare*. Port Townsend, WA, Loompanics Unlimited, 1986.

Royal United Services Institute for Defense Studies. *Nuclear Attack: Civil Defense Aspects Of Civil Defense In The Nuclear Age*. Fairview Park, Elmsford, NY, Brasseys, 1985.

Spigarelli, Jack A. *Crisis Preparedness Handbook.* Provo, UT, Resource Publications, 1984.

Taylor, Chuck. *The Combat Shotgun And Submachine Gun*. Boulder, CO, Paladin Press, 1985.

Taylor, Chuck. *The Complete Book Of Combat Handgunning*. Boulder, CO, Paladin Press, 1982.

US Army's Special Forces Medical Handbook (reprint). Boulder, CO, Paladin Press, 1982.

"Yes, there are books about the skills of apocalypse -- spying, surveillance, fraud, wire-tapping, smuggling, self-defense, lockpicking, gunmanship, eavesdropping, car chasing, civil warfare, surviving jail, and dropping out of sight. Apparently writing books is the way mercenaries bring in spare cash between wars. The books are useful, and it's good the information is freely available (and they definitely inspire interesting dreams), but their advice should be taken with a salt shaker or two and all your wits. A few of these volumes are truly scary. Loompanics is the best of the Libertarian suppliers who carry them. Though full of 'you'll-wish-you'd-read-these-when-it's-too-late' rhetoric, their catalog is genuinely informative."

-THE NEXT WHOLE EARTH CATALOG

Now available:
THE BEST BOOK CATALOG IN THE WORLD!!!

- *Large 8½ x 11 size!*
- *More than 500 of the most controversial and unusual books ever printed!!!*
- *YOU can order EVERY book listed!!!*
- *Periodic Supplements to keep you posted on the LATEST titles available!!!*

We offer hard-to-find books on the world's most unusual subjects. Here are a few of the topics covered IN DEPTH in our exciting new catalog:

- *Hiding/concealment of physical objects! A complete section of the best books ever written on hiding things!*
- *Fake ID/Alternate Identities! The most comprehensive selection of books on this little-known subject ever offered for sale! You have to see it to believe it!*
- *Investigative/Undercover methods and techniques! Professional secrets known only to a few, now revealed for YOU to use! Actual police manuals on shadowing and surveillance!*
- *And much, much more, including Locks and Locksmithing, Self Defense, Intelligence Increase, Life Extension, Money-Making Opportunities, and much, much more!*

Our book catalog is truly THE BEST BOOK CATALOG IN THE WORLD! Order yours today -- you will be very pleased, we know.

(Our catalog is free with the order of any book on the previous page -- or is $2.00 if ordered by itself.)

Loompanics Unlimited
PO Box 1197
Pt Townsend, WA 98368
USA